Ocean Geopolitics
Marine Resources, Maritime Boundary Disputes and the Law of the Sea

Andreas Østhagen

Senior Researcher, Fridtjof Nansen Institute, Norway

Edward Elgar
PUBLISHING

Cheltenham, UK • Northampton, MA, USA

Published by
Edward Elgar Publishing Limited
The Lypiatts
15 Lansdown Road
Cheltenham
Glos GL50 2JA
UK

Edward Elgar Publishing, Inc.
William Pratt House
9 Dewey Court
Northampton
Massachusetts 01060
USA

A catalogue record for this book
is available from the British Library

Library of Congress Control Number: 2022934694

This book is available electronically in the **Elgar**online
Geography, Planning and Tourism subject collection
http://dx.doi.org/10.4337/9781802201567

ISBN 978 1 80220 155 0 (cased)
ISBN 978 1 80220 156 7 (eBook)

Printed and bound by CPI Group (UK) Ltd, Croydon, CR0 4YY

Contents

Figures

Preface

This book is predominantly original, unpublished, independent work by the author, Andreas Østhagen. Small portions of Chapters 5 (Canada) and 7 (Norway) were published in a slightly different frame and format in the article: 'Why Does Canada Have So Many Unresolved Maritime Boundary Disputes?' by Michael Byers and Andreas Østhagen (2017) in *Canadian Yearbook of International Law* 54 (October): 1–62; Permission has been obtained from the co-author to utilise the relevant sections. The co-author, Michael Byers, helped develop some of the ideas incorporated in these sections, but I am the author for the text itself.

Further, minor parts of Chapter 7 (Norway) are based on previously published work by Andreas Østhagen, namely: 'High North, Low Politics Maritime Cooperation with Russia in the Arctic' (2016) in *Arctic Review on Law and Politics* 7 (1): 83–100; 2018; and 'Managing Conflict at Sea: The Case of Norway and Russia in the Svalbard Zone' (2018) in *Arctic Review on Law and Politics* 9: 100–123.

Acknowledgements

There are a number of people that deserve my acknowledgement and gratitude, having helped me along the path to complete this book. First and foremost, I must thank Michael Byers at the University of British Columbia (UBC) for having convinced me to venture across the world (literally) to pursue a degree working on this topic. Having first met in the context of Arctic studies, Michael convinced me to expand my interests and grapple with something novel, of which this book is the product. I could have continued to write about the political changes taking place in the Arctic. Instead, with the assistance of Michael, I dove into Law of the Sea, maritime boundaries, and regional contexts. For the guidance and feedback, I am forever grateful.

When I left the London School of Economics in 2010 having completed a degree, I vowed to never return to academia. Continuing with a doctorate was out of the question. Alas, never say never. While working in Brussels on EU/Arctic affairs, Kristine Offerdal was the first person who included me in an academic project and guided me towards becoming a researcher. For that, I am thankful. Eventually leaving the hallways of bureaucracy and lobbying at the end of 2013, my new colleagues at the Norwegian Institute for Defence Studies—especially Paal Sigurd Hilde, Håkon Lunde Saxi, Robin Allers and Rolf Tamnes—helped me develop as a young researcher.

Returning to studies at UBC in autumn 2015 took some adjustment. Fellow students in Vancouver helped keep it light and fun, which cannot be underestimated with all the seriousness surrounding academia. A special thanks goes to the three I started with; Guðrún Rós Árnadóttir, Hema Nadarajah and Gregor Sharp. In addition to Michael Byers, several academics at UBC helped me develop this thesis topic and gave me valuable insights, especially Richard Price, Katharina Coleman, Antje Ellermann, Lisa Sundstrom and Alan Jacobs, in addition to my committee members Brian Job and Philippe Le Billon.

Moreover, my connection to Bodø, my hometown in North Norway, has been kept alive through a part-time affiliation with the High North Center at Nord University Business School. Frode Mellemvik at the Center has been a continuous source of encouragement and backing, regardless of the venture or idea, for which I am very thankful. My other colleagues at the Center and at Nord University help me keep a 'northern' perspective on life, despite residing—for the time being—further south.

Having started at the Fridtjof Nansen Institute in the summer of 2017, my colleagues there have been of tremendous help giving both feedback and encouragement. In particular Svein Vigeland Rottem, Geir Hønneland, Arild Moe, Pål Wilter Skedsmo, Øystein Jensen, Anne-Kristin Jørgensen, Lars Rowe, Tor Håkon Jackson Inderberg, Davor Vidas, Olav Schram Stokke and Claes Lykke Ragner have provided invaluable input, support and tips on how to manage life at a research institute.

Another crucial component of the writing has been the support provided by my fellow colleagues in The Arctic Institute. Having assisted Malte Humpert in setting up this network of Arctic-engaged young scholars back in 2011, the group of other young professionals grappling with the same challenges have been a relief. Andreas Raspotnik in particular has not only given me feedback and listened to my complaints more than anyone else; he has also become one of my closest friends in the process.

Others have provided treasured input and inspiration during short work stints in North America: Heather Conley during my time at the Center for Strategic and International Studies in Washington DC; Sarah French Rooke during my time at the Munk-Gordon Arctic Security Program in Toronto; and Oran Young during my time at the University of California: Santa Barbara. Moreover, several individuals have taken the time to meet with me over the years and have provided support or insights that has helped this book come into being. These include, amongst others, Áslaug Ásgeirsdóttir, David Balton, Klaus Dodds, Rolf Einar Fife, Alf Håkon Hoel, Kalevi Holsti, Rob Huebert, Stuart Kaye, Whitney Lackenbauer, Suzanne Lalonde, Ted McDorman, Sara B. Mitchell, Anne Kari Ovind, Tony Penikett, Donald Rothwell, Elana Wilson Rowe, Clive Schofield, David VanderZwaag, and the excellent language editors Susan Høivik and Chris Saunders.

Finally, all the academic and professional support would be futile without the comfort and stability provided by close family and friends. My family has never wavered in their support of my efforts, especially Grethe, Kåre and Vegard. My friends have continued to engage in a topic far from their own everyday lives. Most importantly, my wife, Victoria, has kept up with my writing stress, mood swings and continuous rambling about maritime boundaries. At some point in the process of completing this book manuscript we even managed to bring tiny Theodora into the world. I could not have asked for more love or support.

1.   Sea of troubles

In 2010, Norway and Russia agreed on a maritime boundary in the Arctic, stretching from the Eurasian landmass almost all the way to the North Pole. The new 1,750-kilometre (1,087-mile) boundary was ten times the length of the land border between the two countries and was hailed as a sign of a new 'era' in Norway–Russia relations, as well as Arctic governance more broadly (Lavrov and Støre 2010; Moe, Fjærtoft, and Øverland 2011). Pundits were quick to argue that the primary reason for the maritime boundary agreement must have been the presence of oil and gas resources, not least because resource extraction figured prominently in the two countries' newly launched Arctic strategies (Holsbø 2011).

However, the presence of oil and gas resources does not always prompt an agreement. In the summers of 2019, Cyprus issued arrest warrants for the crew of a Turkish vessel found drilling off the west coast of the island, within what Cyprus deems is *its* exclusive economic zone (EEZ). The US State Department called the actions by Turkey 'highly provocative' and French President Macron 'urged Turkey to stop "illegal activities"' (Vey 2019). Turkish President Erdogan countered: 'The legitimate rights of Turkey and the Turkish Republic of Northern Cyprus over the energy resources of the Eastern Mediterranean are not debatable' (Guggenheim 2019).

These two ocean disputes have very different origins. The Cyprus dispute concerns the conflict over the division of Cyprus and—as an extension—what maritime rights befall the Republic of Cyprus to the south and the Turkish Republic of Northern Cyprus (not internationally recognised) to the north. The dispute between Norway and Russia concerned the location of a maritime boundary from the border of the two countries on the Eurasian mainland stretching northwards into the Barents Sea.

Nevertheless, the Cyprus maritime dispute in 2019 concerns the same issue resolved between Norway and Russia in 2010: the delimitation of sovereign rights and space at sea. What drove Norway and Russia to an agreement in 2010? Why was it not settled in 1977, when the dispute arose, or the late 1990s when negotiations were restarted? Indeed, why is it not, like the case of Cyprus–Turkey, still in dispute?

It is unlikely that Norway and Russia would have been able to reach an arrangement today, more than a decade after the boundary was settled. As put by a former Norwegian foreign minister explaining one of the factors behind the agreement: 'There must be trust between the negotiating partners' (Støre 2010). The worsening in relations between the two countries after the Russian annexation of Ukraine in 2014 have made bilateral relations resemble those of the Cold War, when the two countries were on opposing sides in the larger East–West dispute.

This speaks to the challenge of settling disputes at sea. Currently, across all continents, almost half of all maritime boundaries are still disputed. Moreover, maritime boundary disputes exist on all continents: an overview provided in this book shows that in 2020, out of the 460 possible maritime boundaries, only 280 have been agreed upon. This leaves 180 outstanding maritime boundary disputes, or approximately 39%.

As put by the Norwegian and Russian foreign ministers in 2010: 'unresolved maritime boundaries can be among the most difficult disputes for states to resolve' (Lavrov and Støre 2010). What has hindered a conflict from arising between these two countries that have historically been adversaries? In the case of Cyprus, what might lead the different actors engaged—including Turkey and the European Union (EU)—to reach a compromise? Is a mutually beneficial arrangement even possible? Or are the parties best served by leaving the dispute unresolved for the foreseeable future? To answer these questions, we must examine who owns the ocean by looking at disputes over maritime boundaries and how these link to international politics—that is, geopolitics—over ocean space.

1.1   THE CHANGING GEOPOLITICS OF THE OCEANS

The oceans are the foundation of human life: as a point of origin for the human species, as an essential source of food and oxygen and as the constant regulator of global climate. At the same time, the oceans are inaccessible and capricious, a domain never permanently conquered or inhabited—apart from, perhaps, Plato's lost civilisation of Atlantis. The oceans have also been central in the development of civilisations (i.e., Paine 2013; Steinberg 1999); they have enabled the rapid movement of people, goods and ideas among countries and between continents.

The difference between land and sea is clear; and the maritime domain has been kept separate from land in our attempts at explaining political and economic development.[1] How states have viewed and utilised the sea—eventually attempting to control and develop a legal order for it—has varied and changed over the past millennium (Benton 2010, 153–54; Steinberg 2001). From the fifteenth to nineteenth centuries, the use of maritime space in exploration, dominance and industrialisation transformed the world (e.g., Paine 2013). Since the end of the Second World War, the increase in global trade has had a considerable effect on the use of oceans. World trade has grown in real terms from 0.45 trillion constant USD in the early 1960s to 16 trillion in 2017 (Bernhofen, El-Sahli, and Kneller 2016, 36; WTO 2019, 5). This increase is mainly due to the introduction of 'containerisation' in the 1950s: the use of containers to move goods from port to port with cargo vessels.

Today, some 80% of all global trade and 70% of the value of this trade is transported by sea by a global fleet of some 58,000 ships (Devabhaktuni and Kennedy 2012, x). The oceans have always been a base for resource extraction, although marine resources are becoming increasingly relevant in the global context. World per capita fish consumption, for example, is rising twice as fast (3.2%) as population growth (1.6%), from an average of 9.9 kg in the 1960s to preliminary estimates indicating growth beyond 20 kg (FAO 2016, 3–5; FAO 2018, 2).

The effects of climate change on the oceans have also become increasingly apparent in recent decades. Specific issues such as plastic pollution in the oceans have received considerable focus from the media and politicians alike (Harrabin 2017).[2] According to the Intergovernmental Panel on Climate Change, sea levels might rise by *at least* one metre by the year 2100 (IPCC 2013, 25). That, in turn, may influence the question of who owns what at sea: with changes in the baselines from which boundaries or in the characteristics of islands and territory, states may find themselves faced with new challenges or be forced to revisit old and unresolved disputes (Caron 2009, 12–13; Árnadóttir 2016). This could cause further tension, even conflict, among states (e.g., Rayfuse 2009; Lusthaus 2010).

Ongoing changes in the oceans further affect states' access and rights to marine resources, which are expected to become increasingly scarce, like fisheries (Pauly and Zeller 2016; Economist 2017d). Most marine living resources are transnational in distribution, forcing states to interact regarding their management or to pursue a delimitation of rights and access (Prescott and Schofield 2005, 216). With water temperature and currents changing, conflicts over who gets to fish what and where are likely to increase (Pomeroy et al. 2007; VanderZwaag 2010; Spijkers et al. 2018).

Linked to this, there has been a widespread reduction in the total biomass of marine resources, predominantly due to human exploitative activities (FAO 2014, iiv–iv, 3–9; Russell 2010). According to the United Nation's (UN) Food and Agriculture Organization (FAO), at least 32% of fish stocks are overexploited (Economist 2017d; FAO 2016). In fact, these figures are likely too low, given under-reporting and inaccurate assessments (Wood et al. 2008; Pauly and Zeller 2016).

Accordingly, we have observed rapid changes in our oceans over the past few decades. Changes deriving from resource pressures, international commodity prices and new technology are external to the ocean. Rising sea levels and other changes in the oceans resulting from climate change and changing resource distributions are happening *in* the ocean and, to varying degrees, are the consequences of human behaviour.

Particularly ripe for conflict between states with opposing views on how to delineate and manage the ocean are those regions affected by both types of change—like those with great economic potential, even as rapid changes are underway in the ocean itself. This is the changing *geopolitics* of ocean space.

In the South China Sea, for example, disputes involving China, Vietnam, the Philippines, Brunei, Taiwan and Malaysia have escalated in recent years, as power relations and other political dynamics change and marine resources grow in importance and scarcity (Kaplan 2011; Simon 2012; Emmers 2010; Nasu and Rothwell 2014). Similarly, in the Arctic Ocean, the melting of the sea ice and heightened attention paid to the region have led to a focus on pre-viously neglected maritime disputes in the North (Rothwell 1996; Hoel 2009; Byers 2013; Østhagen 2018a).

States are reacting to these trends and developments by engaging in inter-national efforts that can deal with the maritime domain. An agreement under the UN Convention on the Law of the Sea (LOS) on the conservation and sustainable use of marine biological diversity in areas beyond national juris-diction is currently under negotiation (Lalonde 2010; Prip 2017). Similarly, a new regulatory framework for seabed mining under the International Seabed Authority (ISA) is being developed.

On the other hand, states are also encouraging and enabling fishing, oil and gas, and mining companies to engage in greater resource extraction or are being pressed by these actors to allow commercial activity in areas previously dismissed as uninteresting or off limits.

All these trends are leading to a new 'era' for maritime issues and maritime space (Hannigan 2017). States have rights and duties regarding maritime space, and as this space gains attention, the delineation of ownership and rights is already rising to the fore of domestic and international politics. As this happens, more and more attention will be paid to the question of 'who owns what' at sea.

1.2   UNPACKING THE COMPLEXITIES OF MARITIME DISPUTES

With the rapid global economic growth across borders and between continents that has taken place since the end of the Cold War, maritime space has been ascending in importance. This is not an immediate change. An examination of older literature, from Greece to Iceland and China, reveals the prominence of the oceans over the past three millennia. As put by Mahan and Beresford at the end of the nineteenth century:

> Control of the sea, by maritime commerce and naval supremacy, means predominant influence in the world; because, however great the wealth product of the land, nothing facilitates the necessary exchanges as does the sea. (1894, 559)

However, it is only recently—in an extended view of history—that states' ability to uphold sovereignty at sea has led to oceans becoming subject to explicit international jurisdiction. When states attempted to legalise the maritime domain in the twentieth century, with the Geneva Conventions on the Law of the Sea in 1958 and the United Nations Convention on the Law of the Sea (LOS) regime in 1982, the relationship between states and ocean space changed.[3]

After the LOS was adopted in 1982 and states began claiming expansive maritime zones, many boundary disputes arose. Today, maritime disputes over where to draw boundaries at sea exist on all continents—and almost half of all maritime boundaries remain unsettled (Ásgeirsdóttir and Steinwand 2016, 10; Østhagen 2021b). On the one hand, scholars have approached these disputes from a technical or legal point of view.[4] As D. M. Johnston argues, boundary-making in the ocean is functionalist: it is done with an eye towards the *functional usage* of the maritime space itself (D. M. Johnston 1988; D. M. Johnston and Valencia 1991). However, this does not explain why so many maritime boundaries have been left unresolved or why the conflict potential of maritime disputes seems to be on the rise.

On the other hand, scant attention has been paid to the maritime domain as a distinctive site for the study of state behaviour, conflict and international relations (IR).[5] Some scholars have argued that different political patterns derive from maritime disputes than from those on land, such that 'the political salience of the dispute is generally limited, in contrast with the importance and attention often given to land-based disputes' (Huth 1998, 26). Similarly, Hensel et al. (2008, 121, 138–40) hold that maritime territory lacks the 'intangible dimension' that land territory possesses and, therefore, is less likely to reach the top of the political agenda and lead to conflict.

This implies that we are better served by studying territorial disputes on land, and conflict theories in IR have often been developed with a focus on disputed land. Maritime disputes have often been dismissed as peripheral in the conflict literature. The research on maritime disputes tends to be case oriented and is rarely seen in relation to the more general literature on interstate conflicts. Maritime space is either excluded from such studies or presented as apolitical, with the oceans frequently being seen as primarily a resource base, not of existential importance to states (Hensel et al. 2008; Huth 1998; Wiegand 2011a).

Thus, maritime space is *depoliticised*, reduced to technical, economic and legal factors. However, if this truly was the case, we would assume states would not have a hard time agreeing on a boundary. Why do conflicts keep erupting at sea? Why are ocean-based conflicts increasingly on the political agenda?[6]

Historic resource conflicts and contemporary boundary disputes all around the world indicate the range of economic and political interests involved in maritime delimitation. Today's heightened focus on maritime issues is prompting a quest for new approaches to solve the old question of 'who owns what' at sea, calling into question the assumption, which is prevalent in studies of IR and conflict, that maritime space is more prone to dispute resolution—and less to escalation and militarisation—due to its limited salience and intangible importance.[7] As Kaplan (2011, 79) argues, '[t]he sea, unlike land, creates clearly defined borders, giving it the potential to reduce conflict'. However, does it? How? And is the underlying basis for such assumptions stable or changing?

From a political science perspective, the literature on maritime boundary disputes has remained underdeveloped. When politics come into play, the settlement of boundaries is arguably more than just a legal or technical matter. Simply describing maritime disputes as distinct from disputes on land, whose intangible or symbolic value (unlike their material value) tends to be lower, fails to capture the politics involved.

Moreover, some assumptions underlying the current conceptualisation of the maritime domain in international conflict studies—as outlined in Chapters 2 and 3—may be changing, due not in the least to technological progress and the greater political focus on oceans. No studies have systematically explored the *causes* of agreement on *maritime* boundary disputes across cases. Yet maritime space has become increasingly important.

An unsettled maritime boundary can hinder economic exploitation of offshore resources. Similarly, it may complicate the management of transboundary fish stocks. At times, states engage in indirect conflict over such disputes, whether by arresting fishing vessels from the other party or by engaging with navy or coast guard vessels directly. If settling an outstanding maritime dispute is done to serve a function and failure to agree on a boundary can have adverse effects, why, then, are not more boundaries settled? There would seem to be more factors involved than mere function when states consider resolving their maritime boundary disputes.

The notion of a 'boundary' in the ocean that delineates who owns what is in and of itself a somewhat illusive concept. As will be explored in Chapter 2, determining a maritime boundary is inherently a technical process that is usually based on widely accepted legal principles, as merely a line on a map without defined physical markers (in contrast to a border on land). However, because maritime boundaries define the space in which states operate—as do companies and individuals—settling maritime disputes is also a highly political process with potentially far-reaching consequences.[8] It is this *politicisation* of maritime space that prompts a study of how states approach disputes at sea—that is, the geopolitics of ocean politics.

1.3   THE PURPOSE OF THIS BOOK

This book examines developments in international politics and law that enable states to settle maritime boundary disputes, what prevents them from doing so and why this matters in the first place. As the maritime domain has become legalised over the course of the past century, disputes have arose between states, across *all* continents. Exactly why some states managed to agree on a settlement of a maritime boundary at a given point and with what motivation is unclear. We need to understand the mechanisms involved in maritime disputes—across cases and geography. Why do states settle some maritime boundary disputes but not others? Specifically, what factors contribute to a situation where states manage to reach a settlement? Why do some maritime boundary disputes escalate into conflict, whereas others are settled peacefully? Why do some disputes remain unaddressed but unproblematic?

With the answer(s) to these questions come additional questions. How can we explain the processes surrounding maritime dispute resolution, which—according to international law—rests on finding an 'equitable outcome'? Why do many disputes persist and escalate, given the relatively technical and apolitical characteristics of the international law applicable to the maritime domain? How is the relationship between society and the oceans changing? What are the consequences of these changes for the global governance of maritime space and ocean geopolitics? These questions speak to both international law and IR.

This book will demonstrate the necessity of studying maritime boundary disputes in their broader political and legal context. For example, a country with several disputes does not approach each dispute as separate from the remaining ones (and its maritime neighbours): rather, the disputes are viewed as being linked, albeit to varying degrees. This situation of individual disputes in the broader context makes it possible to tease out the factors and mechanisms that are relevant *across* cases.

In sum, this book seeks to explain why some states settle their maritime disputes whereas other disputes escalate into conflict, showing how this can help us understand what changes are currently underway that impact the geopolitics of oceans. Studying how states deal with disputes over the delineation of maritime rights offers insights into the broader relationship between states and the sea—which is of relevance to understanding of IR, state behaviour and spatial domains more generally. Further, this will help us better understand the nuances pertaining to conflict prevention and resource management at sea, at a time when these issues are ascending on the agenda around the globe.

1.4   APPROACH AND FRAMEWORK

Just as there are many ways of conceptualising IR, various theoretical pathways offer different approaches to this topic. Studies have shown that throughout human history, territorial disputes have been the primary source of interstate conflict (Holsti 1991; Zacher 2001).

On the one hand, it is argued that states devote attention and resources to settling disputes only when they become sufficiently politicised and reach the top of the political (and economic) agenda (Hensel et al. 2008). In other words, a dispute will be settled only when it is deemed sufficiently worthwhile. On the other hand, it has been argued that states (and their leaders) manage to settle a dispute only when the 'salience' of the dispute is low and is kept out of the political limelight (Weil 1989; Prescott and Schofield 2005; Wiegand 2011a). In other words, the fact that the dispute does not figure into the political agenda makes settlement possible because high salience makes compromise more difficult to achieve. An additional (and interlinked) explanation is that states will tend to resolve those disputes that are relatively uncomplicated (legally or otherwise) (Byers and Østhagen 2017). How do we evaluate and understand these different political processes and how they are related to how states manage ocean space?

Concepts such as 'salience', 'urgency' and 'value' recur in studies of interstate conflict over issues ranging from territory to resources.[9] A maritime dispute with 'high' scores on all these accounts is more likely to be settled than one with 'low' scores.[10] However, what determines 'salience'? Why do some states perceive a dispute as more 'urgent' than others? What makes a maritime domain 'valuable'? Further, what are the *processes* that link, for example, urgency or value to actual settlement? How do the legal factors come into play here?

This book attempts to break new ground in answering these complex questions through an interdisciplinary approach, drawing primarily on two fields of study: IR (as a subfield of political science) and the LOS (as a subfield of international law). This approach rests on other scholars having ventured across this interdisciplinary divide: lawyers applying IR theory or political scientists making use of international law.[11]

I adhere to a pragmatic approach to theories and theoretical frameworks. Admittedly, mixing too many approaches and explanatory models based on competing ontological and epistemological views can be detrimental to the applicability and leverage of the findings (Hollis and Smith 1996, 112). Still, both a rational and ideational approach to theory will be utilised rather 'eclectically' (Katzenstein and Sil 2008).

Rationalist approaches place emphasis on the goal-seeking *rational* behaviour of states, which are driven and constrained by material factors and institutional structures.[12] Systemic and/or domestic structures constrain state action and induce certain observable patterns of behaviour under specific conditions. Ideationalist approaches take into consideration the *ideational* factors as well.[13] These include the conceptualisation of identity, the role of historical images and the mutually constitutive and socialising processes that occur among actors.[14] The observable patterns and/or causal logics of these processes might be harder—and at times almost impossible—to observe although that does not discount their relevance for the question(s) at hand (Furlong and March 2002, 20).

It is probably unlikely to reveal infallible laws that will determine when states escalate a dispute over maritime space. What we can do, though, is seek to identify the conditions under which it is more or less probable that states will raise a dispute into a conflict. Still, we must also allow room for the intricacies of each case. This demands that we determine the complexity of each dispute on its own, delving beyond initial (and superficial) explanations that hinge on factors general to all disputes.

To strike a balance between the large number of possible cases and the desire to examine a few cases in depth, I employ a two-step approach. First, I present some general information on the topic, here using a comprehensive dataset of all maritime boundary segments across the globe (460 single maritime boundary segments). I draw on a review of the (currently) eight volumes of *International Maritime Boundaries*, which since 1993 has published on all known maritime boundary agreements. That, together with a dataset by Ásgeirsdóttir and Steinwand (2016), which leans on the work done by Pratt and Schofield from 2004, gives a total of 460 maritime boundary segments across the world. As of 2020, 280 of these were agreed, but many agreements are still not ratified; only 25 were settled through arbitration or adjudication. I then offer some propositions regarding what determines the willingness of states to embark on negotiations in the first place and what hampers or drives negotiations forward.

Second, I take four countries and examine their collection of maritime boundaries segment by segment. These countries were selected based on having a large number of maritime boundaries, while excluding extreme outliers (range: more than 5 and less than 15); having boundaries with multiple different countries, ideally in various maritime domains; and not situated in the same region/security environment. Following these criteria, the current study is based on four countries: Australia, Canada, Colombia and Norway.

Having selected these four countries and their related maritime boundaries, I briefly outline the processes and political and legal dynamics in each dispute, as well as how they relate to each country's overarching strategy concerning delineating maritime space. Crucially, I do not view maritime boundary disputes as singular data points and/or outcomes, but rather as part of a larger national and regional complex in which the outcomes and decisions concerning one maritime boundary might have an effect on another involving the same country.

Data come from multiple sources, including public records, secondary literature and—to a limited extent—interviews with former or current officials. This work also draws on international law, with treaties, customary international law and the case law of international courts and tribunals forming the bases on which parts of the examination are undertaken.[15]

One final set of caveats: First, this is a study of the *process* leading up to an agreement on a maritime boundary and the related reasons why an agreement might not have been achieved. It is not a study of the actual maritime boundary line itself or an attempt at explaining why states choose to delineate the *line* as they do. Although considerations of outcomes do figure in the case studies, they are not the primary focus. Further, the cases concern specific boundaries (settled/not settled), but these cannot be disentangled from the country in which they are located. This interplay between the case and national context is crucial, guiding the study. Finally, it should be noted that the primary level of analysis is the state and state-to-state interactions concerning maritime space. This does not, however, discount the role of individuals—such as a prime minister, president or chief negotiator.

1.5   NEXT STEPS

The remainder of this book follows a rather straightforward structure. In Chapter 2, I delve deeper into the foundations of states, territory, borders and boundaries. This provides the empirical and analytical structure underlying this book, in which I showcase the breadth of the phenomenon across contexts and continents. In Chapter 3, I outline how we can understand the state behaviour and international politics/geopolitics concerned with maritime space. This includes political scientists and legal scholars, as well as interdisciplinary work in these fields and from other fields such as geography and economics.

Chapters 4–7 present the empirical analysis, in which I outline in a limited format the different case studies: Australia, Canada, Columbia and Norway. Each subchapter offers a description of the country and the specific cases of maritime boundaries (settled/outstanding), as well as how and why it came to settle its boundaries, where applicable.

Thereafter, Chapters 8, 9 and 10 examine the specific mechanisms and issues of relevance to understanding maritime disputes, conflict at sea and state interests. This starts with international law and how this ever-evolving institutional construct governs ocean geopolitics. Thereafter, I turn to the most obvious driver for both conflict and agreements over maritime space: oil and gas resources. Then, I explore a crucial dimension of maritime boundary disputes and ocean politics more broadly: the link between security interests and fisheries. Finally, Chapter 11 addresses the larger topics of IR, conflict, geopolitics and maritime space.

NOTES

1. This book employs various terms referring to maritime space, such as 'sea', 'ocean', 'maritime domain', 'maritime space' and 'ocean space'. These are, to some extent, used interchangeably, albeit with certain nuances: The Oxford Dictionary defines 'sea' as 'The expanse of salt water that covers most of the earth's surface and surrounds its land masses'. The Oxford Dictionary defines 'ocean' as 'A very large expanse of sea, in particular each of the main areas into which the sea is divided geographically'.
   This distinction is applied here as well: *ocean(s)* refer to a more specific part of the sea (used as a general term for maritime space), whereas a *specific sea* (such as the Mediterranean Sea) denotes a smaller geographic entity than an ocean. *Maritime space* and *ocean space* are used for emphasis on the spatial dimension of the sea, though these terms inherently refer to the same thing.
2. See also Lejeusne et al. 2010; Laffoley and Baxter 2016; Levin and Le Bris 2015.
3. See Osgood 1976; Bull 1976; Brown 1981.
4. See, for example, Friedheim 1993; Brown 1981; Irwin 1980; Bailey 1997; Springer 1994; McDorman 2002.
5. However, there exist area-specific studies of, for example, the Arctic and the South China Sea, as well as literature concerned with defence, sea power and power projection (see, e.g., Booth 1985; Till 2004).
6. See, for example, the *Economist*'s increasing coverage of issues pertaining to the oceans (Economist 2006; 2010; 2011; 2013; 2014b; 2014a; 2016a; 2016b; 2017a; 2017b; 2017c; 2017e).
7. Note the term 'salience', as commonly used in studies of conflict over interstate issues to describe how much political attention a specific issue attracts. See, for example, Hensel et al. 2008; Nemeth et al. 2014; Eiran 2017.
8. Note that 'settled' entails that two states have formally agreed on the exact delineation of a boundary at sea, whether ratified by both countries or as a minimum adhered to as a finalised boundary. When a dispute is 'unsettled', the whole area claimed by both states remains disputed. The process from an unsettled boundary dispute to a formally ratified agreement is further explored in Chapter 3.
9. Linked to the Correlates of War project: http://www.correlatesofwar.org/.
10. See Hensel et al. 2008; Nemeth et al. 2014.
11. See, for example, Byers 2000, 1999a; Reus-Smit 2004; Goldsmith and Posner 2005; Goldstein et al. 2000; Finnemore and Toope 2001; Brunnée and Toope 2010.

12.   See Fearon 1995; Keohane 1984; Keohane and Nye 2012; Krasner 1983; Mearsheimer 2001; Moravcsik 1997; Waltz 1979, 1959.
13.   As outlined by Checkel 2008.
14.   See Adler and Barnett 1998; Campbell 1998; Checkel 1998; Hopf 2002; A. I. Johnston 2001; Wendt 1994, 1999.
15.   Although this volume is not based primarily on interview material, it draws on a range of background interviews conducted over an extended period (2016–2019) to supplement and test the findings acquired through public documents, other academic work and news outlets. I used these interviews to gain additional insights into particularly difficult and sensitive negotiations, confirming the validity of the hypotheses. These interviews were conducted predominantly with government officials and diplomats, who—due to the sensitive nature of international negotiations and the boundary disputes in question—were adamant about remaining anonymous. Where it is possible to do so, I have used full names.

2. States, borders and maritime boundaries

This chapter examines how the concept of territorial ownership and boundaries between communities came about and has been implemented over the past few centuries. Further, it examines how the idea of sovereignty entered the maritime domain, in turn leading to the creation of a legal regime for the oceans. The notion of maritime rights came as a consequence of this 'legalisation' of maritime space, creating the need to delineate maritime zones between states. It is this legal regime that defines the parameters for states when they engage in dispute resolution over maritime boundaries. Finally, this chapter showcases the relevance of maritime (boundary) disputes across continents and countries.

2.1 STATES AND TERRITORY

As a consequence of European state formation and finite territorial space, the concepts of territorial sovereignty and boundaries have come to define the modern state and its relations to others (Krasner 1999; Zacher 2001; Buchanan and Moore 2003). As states formed, developed and expanded, the need to define and uphold territorial boundaries became increasingly relevant (Tilly 1990, 131). As Kratochwil (1986, 32) argues, 'Boundaries are points of contact as well as of separation between a social system and an environment'. In the fourteenth and fifteenth centuries, European states had begun consolidating around permanent military establishments, and the external boundaries of the states became more significant (Tilly 1990, 46). As these states engaged in war, the demand for state-making prompted resource extraction from a given territory. In turn, the focus on taxes created administrative structures that required clearly delineated boundaries. As Tilly (1990, 131) argues, 'Armed men form states ... by defining boundaries, and by exercising jurisdiction within those boundaries'.[1]

A distinction in early boundary setting involves the difference between *frontiers* and *boundaries*. Frontiers were buffers, protectorates or spheres of influence that were seldom clearly defined. Tilly (1990, 70) describes them as resulting from the desire by state leaders to create buffer zones to protect the inner area of their territory. According to Ruggie (1993, 150), '[T]he notion of firm boundary lines between the major territorial formations did not take hold until the thirteenth century; prior to that there were only "frontiers", or large zones of transition'. Kratochwil (1986, 33) holds that the 1659 Treaty of the Pyrenees between France and Spain established the first modern state boundary.

When the emphasis was placed on delimitation of all territory (terrestrial) in the nineteenth and twentieth centuries, these 'frontier' regions became a source of interstate friction because they lacked clear demarcation. Disputes emerged as states sought to expand their territory and define their borders.[2] Even today, related border disputes exist. Some reasons for this are the 'costs of demarcation in an uncharted and hostile environment' (Kratochwil 1986, 37).

In what has become the international system, the concept of territoriality developed slowly. Because of European state formation and the finite territorial space in this part of the world, the concept of territorial sovereignty and boundaries have come to define the modern state and its relations to other states *across* the globe (e.g., Elden 2013; Agnew 1994; Sack 1986). 'The rise of the bounded state as a political unit necessitated a concern with the drawing and redrawing of political borders and the formalization of territorial arrangements' (Storey 2012, 45). Borders (on land) play an integral part in explaining African state formation—or the lack thereof[3]—along with the various state structures developed in Asia.[4]

Disputes emerged—and still emerge—as states sought to expand their territory and define their external boundaries. The link between territory, sovereignty and conflict has been extensively proven (Holsti 1991; Goertz and Diehl 1992; Vasquez 1993; Forsberg 1996; Huth 1998; Wiegand 2011a). The classic territorial dispute involves two states that disagree on where a border should go, either because one state does not recognise another state's border derived from a previously signed treaty or because no treaty exists at all. More complicated disputes concern situations where a state has occupied the territory of another state, where a state does not recognise the sovereignty of another state or where a state does not recognise the independence *and* sovereignty of a seceding state (Huth 1998, 20–23). Vasquez shows how at least 79% of all wars between 1648 and 1990 were fought over territory-related issues (Vasquez 1995).

Krasner (1999) further divides the related concept of (territorial) sovereignty into four ideal types: *legal*, *Westphalian*, *domestic* and *independence sovereignty*. Legal sovereignty involves the practices associated with mutual recognition between states. Most rulers want recognition because it provides them with material and normative resources. Westphalian sovereignty refers to political organisation based on the exclusion of external actors from a given territory. Domestic sovereignty is the organisation of political authority within the state. Finally, interdependence sovereignty concerns the ability of the authorities to regulate cross-border flows (Krasner 1999, 4–7). These various types are not mutually exclusive because some actors may simultaneously hold several types. In turn, Krasner argues that the international embrace of sovereignty is characterised by a pervasive hypocrisy; the international community claims to be upholding sovereign rights, boundaries and responsibilities, but it often violates them in the name of upholding these very conditions.

A dispute over territory and/or sovereignty can be resolved when (1) the occupation of the territory is formally recognised in a treaty or an agreement; (2) an agreement is reached between states over the disputed territory; or (3) the challenger(s) agree(s) to abide by a ruling by the International Court of Justice (ICJ) or another international court or arbitration tribunal.

Scholars agree that boundaries and the integrity of territory constitute a pillar of the modern state-system. As Tilly (1990, 203) argues in his account of European state formation: 'With a few significant exceptions, military conquest across borders has ended, states have ceased fighting each other over disputed territory, and border forces have shifted.' Tracing the development of the norm of 'territorial integrity' in recent centuries, Zacher (2001) shows how the norm has undergone three phases: emergence, acceptance and institutionalisation.[5] Examining all territorial conflicts between 1946 and 2000, he finds that the norm has indeed been commonly accepted through efforts and statements from the 1970s onwards.

However, territorial disputes *still* occur in the international arena. According to Wiegand (2011a), territorial disputes concern 41% of all sovereign states today. Hensel (1999, 137) holds that interstate rivalry is still twice as likely to escalate into war when territory is involved. In a study of 89 ongoing interstate disputes across the world in 2015, 51 were found to involve territory (Oosterveld, De Spiegeleire, and Sweijs 2015, 6). Indeed, territory and where to draw related borders have not lost their importance.

To sum up, territory has been the primary source of conflict between states over the last millennium as states grew into existence, developed and matured. As noted by Weber (1946), it is the monopoly on the use of force in a given *geographical area* that has come to characterise the modern state. The notion of territoriality has come to define the very idea of statehood (Elden 2013).

2.2 MARITIME SPACE AND MARITIME BOUNDARIES

At sea, however, 'territoriality' and the rights of states take on a different form. Inherently a distinct domain altogether, the way in which society has viewed, legalised and utilised the ocean has evolved through history. In the fifteenth century, as European powers pursued colonisation in waters outside Europe, a debate was sparked concerning the status of oceans and what rights nations could have at sea. The ideas of a natural law of nations were retrieved from antiquity and the Middle Ages and used by scholars to argue for various understandings. Grotius became a frequently cited proponent of the right to peaceful commerce and that passage at sea is natural to the 'need of all men to ensure their survival' (Maier 2016, 33). Grotius argued for the freedom of the seas as a way to counter Portuguese and Spanish claims to trade monopolies in the world outside Europe, when they divided the non-Christian world between themselves with the 1494 Treaty of Tordesillas.

The principle of the oceans as global commons came to a boiling point with the idea that nations had rights and sovereignty in nearby waters. For example, Norwegian kings around AD 1000 had claimed sovereignty in the waters adjacent to Norway, stretching all the way to the British shorelines (Theutenberg 1987, 481). In the fifteenth century, a version of this position was advanced by Britain in response to Dutch attempts at dominion of the North Sea. As Maier (2016, 37) describes it:

> The Dutch sent a fishing fleet of two thousand ships protected by an armed squadron to the North Sea waters off the east coast of Britain; and John Selden argued that the ocean's bounty of cod was no more a public good, replenished by nature, than the land, and like the land it could be assigned to particular owners.

Legal scholars like Hugo Grotius (*mare liberum*—freedom of the seas) and John Selden (*mare clausum*—closed seas) have become symbols for the two opposing ways of grappling with questions of maritime ownership and rights. These conceptions of the ocean, which also hold varying degrees of relevance for different maritime spaces (open seas and/or coastal zones), came to dominate the approaches towards the sea in the subsequent centuries, until the international community began negotiating a legal framework for the oceans in the twentieth century.

Already in the eighteenth century, the territorial waters of states were defined as being a 'cannon shot' from land, an idea developed by van Bynkershoek in 1703 and later defined as three nautical miles (n.m.) by Galiami (Anand 1983, 138).[6] The League of Nations attempted to codify international law concerning the oceans in The Hague in 1930 but never managed to reach an agreement (Friedheim 1993).

Then, in 1945, US President Truman declared that the natural resources of the continental shelf were under the exclusive jurisdiction of the coastal state (United States 1945). This rapidly advanced discussions on what rights states have beyond a limited (3 n.m.) territorial sea. Central to the success of this declaration was not only the US position of strength after the Second World War, but also how the principle entitled every coastal state to similar rights and the fact that these sovereign rights did not depend on occupation (Byers 1999b, 91–92). This was later codified in the 1958 Geneva Convention on the Continental Shelf, which preserved the prospect of exclusive coastal state jurisdiction over offshore seabed resources (United Nations 1958).

When states began expanding their maritime zones, the notion of straight baselines—the line drawn along the coast from which the seaward limits are measured—also came to the fore. Instead of drawing the baseline of a country's maritime zone along its coast following all features, some states with indented coastlines or with multiple fringing islands started to draw *straight lines* along the coast, in essence claiming more maritime space (territorial sea) than a country with an even coastline. The UK took a case against Norway to the ICJ concerning this practice, which in 1951 endorsed the Norwegian approach regarding straight baselines with the *Anglo-Norwegian fisheries case* (Green 1952).

At the same time, some states started expanding their territorial seas from three to twelve n.m., as negotiations of an international regime for the oceans were underway. This led to conflict around adjacent and overlapping maritime spaces (Harrison 2011). The first and second Law of the Sea Conferences were held in 1956–1958 and 1960, without reaching final agreement on the extent of the territorial sea or the extension of state rights and jurisdiction extending further offshore beyond the territorial sea (Anand 1983). Then followed decades of negotiations aimed at developing a coherent international legal framework for the oceans; in 1982, most states agreed on a comprehensive legal regime: the United Nations Convention on LOS (UN General Assembly 1982).

When it was agreed upon, the LOS provided the legal rationale for states to implement new maritime zones, in addition to the 12 n.m. territorial sea, with a 200 n.m. 'resource' or 'fisheries' zone (what became termed the EEZ), driven largely by growing awareness of the possibilities for marine natural resource extraction (hydrocarbons, fisheries, minerals) and the desire of states to secure potential future gains (Brown 1981; Friedheim 1993).

Already in 1952, Peru, Chile and Ecuador had made claims of exclusive rights out to 200 n.m., seeking to reap the benefits of an expansion in fisheries (Chile, Ecuador, and Peru 1952). These initial claims wetted the appetite of many coastal states, and afterwards, a diversity of claims were put forward, with other states also claiming resource zones, including exclusive fishery zones in the 1950s, 1960s and 1970s. The international community agreed on the legal regime of the EEZ as defined under Part V of LOS.

As a result, in the span of a few decades, states had gone from having control over a relatively limited (often just 3 n.m.) maritime domain to having an international agreement on expanding the length of the territorial sea where states have full sovereignty to a maximum of 12 n.m., while also adding an EEZ where states have certain sovereign rights for an additional 188 n.m.

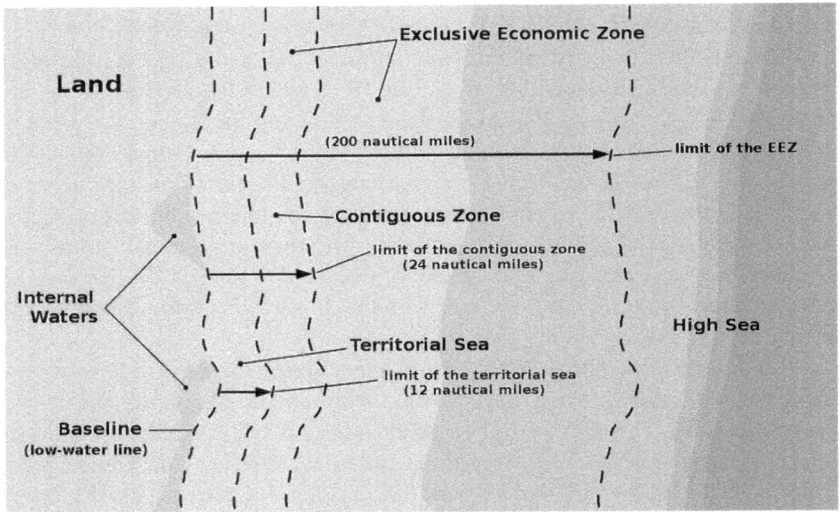

Note: An overview of the different maritime zones awarded to states by the LOS. Note that the continental shelf is not included here.
Source: https://upload.wikimedia.org/wikipedia/commons/c/c8/UNCLOS-en.png.

Figure 2.1   The maritime zones of a state under the LOS

Moreover, with the LOS, it was concluded that states have sovereign rights on the continental shelf up to 200 n.m. and, when relevant and by submitting this information on the limits to the UN Commission on the Limits of the Continental Shelf (CLCS), beyond 200 n.m. where the shelf is a prolongation from the land mass of the coastal state (UN General Assembly 1982) – see Figure 2.1. The limit of such claims was determined to be up to 350 n.m. from a country's baseline or not exceeding 100 n.m. beyond the point where the seabed is at 2,500-metre depth (2,500-metre isobath) (Busch 2018, 321).

With 168 ratifications as of 2021, the LOS has become part of the larger framework of international politics and law (Finnemore and Toope 2001; United Nations 2021). Many of its provisions today reflect customary international law, which is universally binding on all states and not limited to LOS parties only (Roach 2014). This legal–political regime that took decades to develop has enabled states to reach a relative agreement on how to tackle the issues that first arose centuries ago. As Keohane and Nye (1977, 56) put it: 'There is very little direct functional relationship between fishing rights of coastal and distant-water states and rules for access to deep-water minerals on the seabed; yet in conference diplomacy they were increasingly linked together as oceans policy issues.'

However, a central bone of contention that remained—and remains—is how and where to *delineate* maritime space and the related rights to resources on the seabed and in the water column.

2.3   MARITIME BOUNDARY DISPUTES—EQUITY AND EQUIDISTANCE

As states expanded their maritime zones, a number of maritime boundary disputes between neighbouring states emerged. Different states have developed different interpretations of how to draw boundary lines at sea (Forbes 1995); these relate to which map projection to use when drawing the boundary; whether or not to base the boundary on a median principle or a sector principle; the shape of the geographical attributes of the land from which the maritime boundary is derived, that is, the direction of the coastal front and the weight given to islands and submarine features; and which portion of the coast is relevant to delimitation (Bailey 1997; Bateman 2007; Nemeth et al. 2014).

When states expanded their fisheries zones or EEZs to 200 n.m., existing maritime boundary disputes were enlarged as the disputed areas grew in size. Boundary disputes also arose or became more significant between the maritime zones of 'adjacent' or 'opposing' coastal states. Some of these boundary disputes were settled immediately, but a large number remain today. Figure 2.2 displays how the EEZs of countries bundled together are contiguous and, thus, also need a clear boundary.

As maritime zones and state interest in them rose within the political agendas in the middle of the twentieth century and the need for their delimitation increased, the concept of 'equidistance' came to the fore. This guiding principle encountered another principle: equity. The balance between these two principles has shifted over the last half-century, and this tension is crucial in understanding how states settle their maritime boundary disputes (and the principles that guide such processes).

Equidistance entails a boundary that corresponds with the median line at an equal distance (equidistance) at every point from each state's shoreline. Some scholars have taken the position that this was codified under Article 6 (2) of the 1958 Geneva Convention on the Continental Shelf (Geneva Convention), which directs states to settle overlapping claims by reference to the equidistance principle (Franck 1995, 62). As St-Louis (2014, 26) points out, with the Geneva Convention, states 'intended to have equidistance applied as the basic principle, to be deviated from only in the case of special circumstances'.

However, international law is not a static set of rules but rather a process that evolves through time (Byers 1999b). The attention given to 'relevant' or 'special' circumstances led to varying interpretations among states. In addition to coastal length and other geographical variables, security interests and the

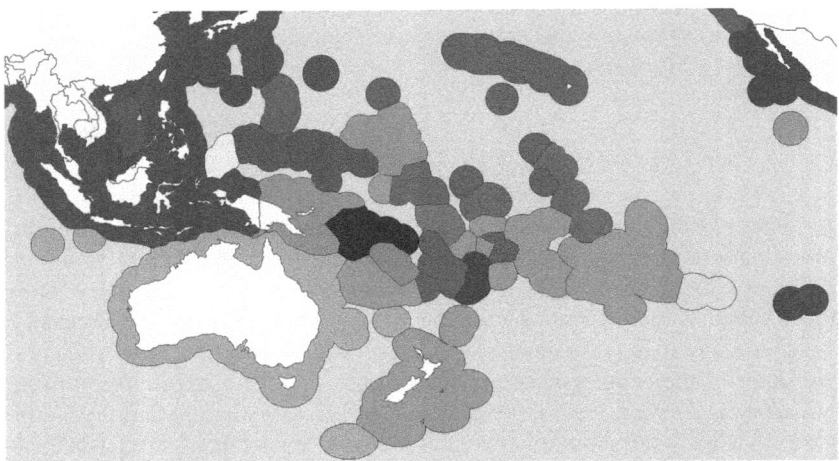

Note: Displaying the numerous EEZs in the South Pacific and how these are adjacent/
overlapping and, thus, have been in need of delimitation.
Source: https://upload.wikimedia.org/wikipedia/commons/thumb/b/b1/EEZ_Oceania.svg/2000px
-EEZ_Oceania.svg.png.

Figure 2.2   The exclusive economic zones in Oceania/the South Pacific

location of natural resources have sometimes been accorded weight in a few
international court rulings. This has been termed 'equity', a principle distinct
from 'equidistance'. In general, equity provides one of the foundations for
national law, as well as one of 'the general principles of law recognized by
civilized nations' (United Nations 1946, Art. 38 ICJ statute). Equity is often
coupled with or explained as 'fairness' (Franck and Sughrue 1993, 564). As
Haywood Jefferson Powell (1993, 8) states it: 'Law … is suffused with tradi-
tionally equitable invocations of fairness and conscientious behavior.'

Equity acquired importance in delimiting disputes the maritime domain
(Cottier 2015). In particular, the North Sea Continental Shelf cases between
Denmark, West Germany and the Netherlands from 1969 pitted the principle
of equity and equidistance against each other (Oude Elferink 2013). Denmark
and the Netherlands argued for the use of equidistance, whereas West
Germany argued for a 'just and equitable share' of the disputed area. Outlining
its approach to maritime boundary dispute settlement in general, the court held
that delimitation must be 'effected in accordance with equitable principles …
taking account of all the relevant circumstances' (ICJ 1969, 53).

In addition, the court introduced the concept of the 'natural prolongation'
of the continental shelf—that also the geophysical attributes of the shelf in
question matter for delineation between states (Kaye 2001, 15). Although the
ICJ specified that there was 'no legal limit' to the number of factors relevant

to delimitation of the shelf, these were initially defined as geology, the desirability of maintaining unity of the natural resource deposits and proportionality (the ratio between the water and shelf areas attributed to each state and the length of their coastline) (ICJ 1969, 51–52).

Thus, states were not deemed as being obliged to apply the equidistance principle: equity was seen as extending beyond mere equidistance (Oude Elferink 2013; St-Louis 2014). Robert Kolb (2003, 108) argues that the ICJ's rulings in the 1960s and 1970s changed the jurisprudence from method (equidistance) to objective (equity). This entails that not equidistance—but fairness on its own—was introduced as a guiding principle for maritime dispute resolution.

In the 1977 Anglo-French Continental Shelf case, for example, the ICJ included factors well beyond equidistance. As Thomas Franck (1995, 64) argues in *Fairness in International Law and Institutions*:

> In interpreting 'special circumstances' to include such factors as the islands' populousness and political and economic importance as well as defense considerations, the arbitrators were able to claim that they were applying normative principles of justice. To protect against allegations of unbridled subjectivity, they brandished their overall reliance on the equidistance rule, from which departure was made only to take into account special circumstances.

Scholars have outlined how in the late 1970s LOS negotiations concerning maritime boundary dispute resolution reached a compromise between two groups of states: those wanting the equidistance principle enshrined and those wanting equity as the guiding principle without specifying any particular method (Brown 1981; Bailey 1997; Kaye 2001; Rothwell and Stephens 2016). Equity as a principle was incorporated in 1982 in LOS Article 74 (*delimitation of the exclusive economic zone*) and Article 83 (*delimitation of the continental shelf*), with the wording, 'The delimitation of the exclusive economic zone/continental shelf between States with opposite or adjacent coasts shall be effected by agreement on the basis of international law … in order to achieve an equitable solution' (UN General Assembly 1982). Kaye argues that 'The result was an acceptable (if fragile) compromise, but one that did little to clarify the method by which delimitation was to take place' (2001, 16). Consequently, the LOS regime does not specify *how* states are to settle maritime boundary disputes—it merely calls for 'an equitable solution' (UN General Assembly 1982).[7]

For example, in 1980, Denmark extended its 200-mile fisheries zone northwards along the east coast of Greenland (Denmark being the colonial power operating on behalf of Greenland), creating an overlap with the Norwegian zone on the northwest side of the island of Jan Mayen (Churchill 2001). Denmark argued that it deserved a larger proportion of this disputed zone because Greenland's coast is longer than that of Jan Mayen and because the population of Greenland deserved privileged access to fish stocks (Churchill 1994). Norway held firm to the equidistance principle; after years of unsuccessful negotiations, Denmark submitted the dispute to the ICJ in 1988.

The court concluded that the longer length of the Greenland coast required a delimitation that tracked closer to Jan Mayen (Maritime Delimitation in the Area between Greenland and Jan Mayen (Denmark v. Norway) 1993) and that the maritime boundary line should be shifted somewhat eastwards to allow Greenland equitable access to fish stocks (Hoel 2014, 55; Churchill 1994). However, the court rejected other arguments concerning population size and socio-economic conditions, declaring them irrelevant to the final determination of the boundary line. Further, even if the court appeared to view ice in the disputed area as a potential relevant/special circumstance, they rejected that as well in this case (Churchill 1994, 8).

As these cases and developments show, how states initially divided maritime space among themselves became questions where maritime law rested on the principles of both equity and equidistance. Thus, the process of settling a maritime boundary has not been very straightforward. As Finnemore and Toope put it (2001, 748):

> If one considers the decisions of the International Court of Justice in boundary delimitation cases, for example, the results are clearly legal, influential, and effective in promoting compliance, but they are highly imprecise.

The dissenting opinion of Judge Gros in the Gulf of Maine case asserts '... equity left, without any objective elements of control, to the wisdom of the judge reminds us that equity was once measured by "the Chancellor's foot"'[8] (Harris and Sivakumaran 2015, section 2-045). Similarly, Brownlie (1980, 288) argues, '... as a general reservoir of ideas and solutions for sophisticated problems it offers little but disappointment'. Equity as a guiding principle found its way into maritime delimitation, albeit without a clear operating definition, and has for that reason been criticised for being a 'unduly subjective and uncertain element' (Harris and Sivakumaran 2015, section 2-045).

However, in rulings in recent decades, the ICJ has favoured a stricter interpretation of which relevant circumstances to include. In the case of Tunisia v. Libya in 1982, the ICJ explicitly rejected equity-based claims founded on relative poverty and redistribution (Harris and Sivakumaran 2015, section 7-210). Instead, it has placed emphasis on geographical factors in a three-stage approach in delineating maritime boundaries (Oude Elferink, Henriksen, and Busch 2018a, 381), as later outlined in the Black Sea case between Romania and Ukraine in 2009 (ICJ 2009). First, a 'provisional delimitation line' between the disputing countries is established, based on equidistance. Second, consideration is given to of 'relevant circumstances' that might require an adjustment of this line to achieve an 'equitable result'. This is where 'equity' is considered. Third, the court evaluates whether the provisional line would entail any 'marked disproportion', taking the coastal lengths of the states into consideration (ICJ 2009, paras. 116–122).

Equity in terms of maritime boundary disputes has become operationalised as a principle within the second and third steps, which is subject to the judges' discretion and where more than just geographical facts are taken into consideration. What constitutes equity has become a highly complex topic, subject to the states and those acting on their behalf, as well as international judges. As St-Louis (2014, 50) states: '[E]quity is defined, not in the context of justice, but in the context of what is just'. The widening of the concept to include societal and political factors—feared by some, applauded by others—has, however, been slightly abandoned to ensure greater consistency in maritime law (Harrison 2011).

Nevertheless, 'while equidistance seemingly has moved centre-stage, relevant circumstances continue to exercise an influence in the background' (Oude Elferink, Henriksen, and Busch 2018a, 381). That being said, law is not in stasis, and new challenges deriving from physical, political, technological and economic changes have entered the picture.[9]

Concerning the continental shelf vis-à-vis the EEZ, initially, the process to settle the two kinds of boundaries were different. The emphasis on 'natural prolongation' and the scientific elements to prove it involved using a different approach for continental shelf delimitation. However, as state practice and court rulings developed after the 1969 North Sea Continental Shelf cases, the principle of natural prolongation lost its hold. The main reason was the introduction of the 200 n.m. concept, where states, regardless of submarine features, immediately acquired rights over the seabed and water column out to 200 n.m. from shore. With the new rules in the LOS and the move away from 'natural prolongation' as a basis of entitlement to the continental shelf, the courts have adopted a uniform approach to maritime boundary delimitation for both the water column and seabed.

This does not mean, however, that the idea of natural prolongation has become totally irrelevant. As Kaye (2001, 19) argues, it has relevance if the submarine feature is 'vast and significant'. Further, the notion of natural prolongation has remained the determining factor concerning 'extended' continental shelves because states must use scientific data concerning the seabed in its submission to the CLCS. Then, the geomorphology (and to a lesser extent, the geology) of the seabed and the ability of states to prove their natural extension come into play. In Chapters 9 and 11, I return to this growing interest in extended seabed claims; let me simply note that the primary focus here is on the delineation of maritime space that includes both the seabed and the water column, and not—except in a few instances—the extended continental shelf.

However, the findings and results presented here do have a bearing on that process as well.

In summary, the concept of boundary-making at sea is in itself based on abstract lines on the map, not borders that physically separate the maritime domains of two countries. States have leaned on—albeit often deviating from—the legal principles set out in international court rulings and the LOS as they attempt to agree on how to draw lines at sea (Oude Elferink, Henriksen, and Busch 2018b).

2.4   A WORLDWIDE PHENOMENON

When a maritime boundary is to be drawn between two disputing nations, finding what can be deemed the equitable solution seems to constitute the heart of the problem—a problem of both a *legal* and *political* character. As put by legal scholar Lasswell (1936), politics is a matter of 'who gets what, when and how'. Many maritime boundaries have been left unresolved for decades because of incompatible and entrenched legal positions and/or limited political and commercial interest (Prescott and Schofield 2005). The legal regime that took shape over the course of the past century came to define precisely what rights states could claim at sea, so processes were developed for settling the disputes that inevitably arose. Because these hinge on political considerations and context beyond purely legal principles and/or geography, we need to employ a wider lens in exploring maritime boundaries.

The LOS (Part XV) encourages states to reach an agreement among themselves after bilateral (or broader) negotiations. If this is not possible or not of interest, states may choose among various pathways for settling disputes: they may submit the case for adjudication at the ICJ or another international court such as the International Tribunal for the Law of the Sea (ITLOS), or they may use third-party arbitration, such as the Permanent Court of Arbitration (PCA). International arbitration is, however, generally unappealing. Uncertainty regarding the outcome of international adjudication and arbitration does not inspire states to bring cases before courts and tribunals. Resolving a dispute bilaterally leaves states with the option of a creative resolution that is not confined by the international rules applied by courts and tribunals (D. M. Johnston 1988, 14–15). Moreover, litigation is costly, and in the maritime domain, the process often requires a great deal of scientific data, making it expensive for states to pursue delimitation in this manner (Prescott and Schofield 2005, 245).

Consequently, approximately 95% of those maritime boundaries that have been agreed upon between 1950 and 2021 were settled through negotiations outside the realms of arbitration or adjudication, with states free to choose whichever approach they prefer when delineating maritime space. However, studies show that although states choose bilateral negotiations to avoid the

Table 2.1   *Total number of maritime boundaries as dyads*

Total number of boundary segments	**460**
Number of settled boundary segments, by 2021	280
- settled through adjudication / arbitration	25
- ratified	243
Remaining boundaries in dispute, by 2021	**180**

Note: Based on the dataset by Andreas Østhagen.

Table 2.2   *Settled/not settled maritime boundaries as dyads across continents*

Continent	Boundaries	Agreements	Still in dispute	Settlement rate (%)
Africa	92	32	60	35
Asia	102	62	40	61
Europe	97	79	18	81
North America	89	45	44	51
Oceania	50	37	13	74
South America	30	25	5	83
Total	**460**	**280**	**180**	61

Note: Based on the dataset by Andreas Østhagen. In some instances, a dyad (consisting of one boundary segment with two countries) may consist of countries from two different continents (e.g., Africa and Europe in the Mediterranean). In such instances, allocations to a specific continent were made on a case-by-case basis.

shackles of international adjudication and arbitration, they still lean on—and generally adhere to—the legal principles as set out by international court rulings (Nemeth et al. 2014; Ásgeirsdóttir and Steinwand 2015; Qiu and Gullett 2017; Østhagen 2020; Lalonde 2002). Still, maritime boundary disputes exist on all continents, ranging from active and conflictual to dormant or successfully managed. Table 2.1 displays the total number of maritime boundaries as dyads between two states, settled or still in dispute per 2021. Table 2.2 shows the settlement rate across various continents, whereas Figure 2.3 highlights the global relevance of maritime boundaries, marking those with the highest number of boundaries (settled and unsettled) more shaded.

Perhaps more interesting are the countries with *outstanding* disputes, as highlighted in Figure 2.4, as well as the ratio settled/unsettled per continent (Figure 2.5).

These figures give a rough idea of the global extent of this phenomenon, which is not confined to one part of the world or a specific group of states.

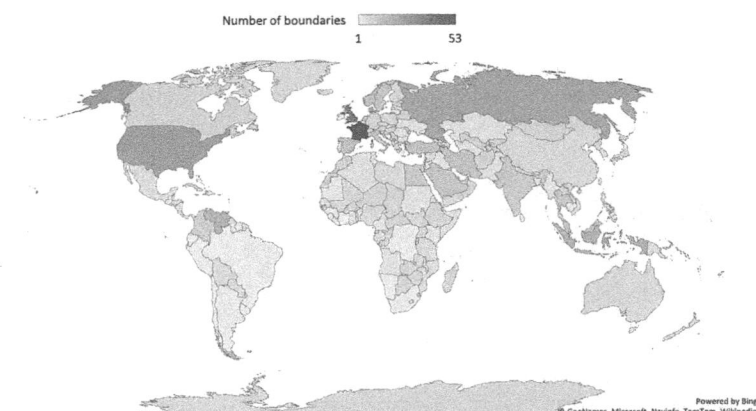

Note: Altogether 161 countries and 480 bilateral boundary segments, ranging from 1 to 53 boundaries per country.

Figure 2.3   Total number of countries with maritime boundaries

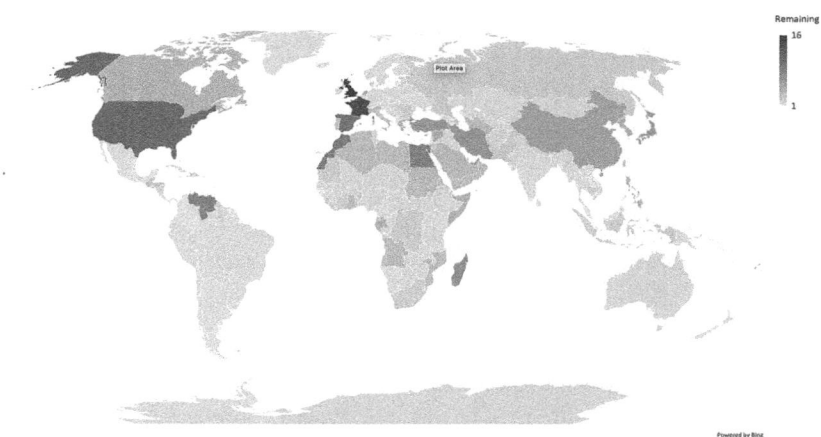

Note: Altogether 121 countries and 180 bilateral boundary segments, ranging from 1 to 16 boundaries per country.

Figure 2.4   Countries with remaining maritime boundary disputes per 2021

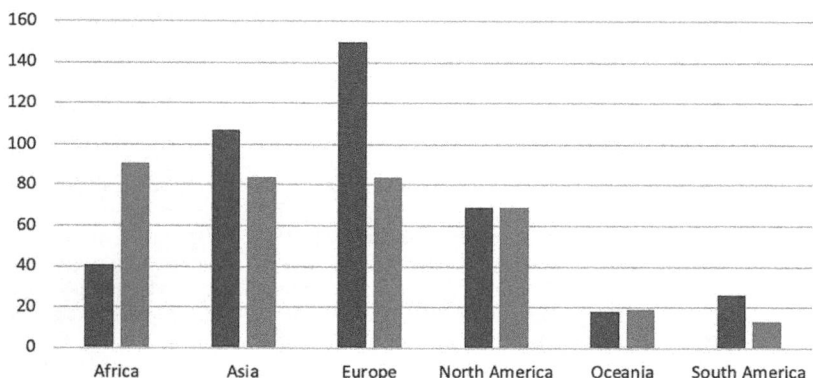

Note: Darker tint denotes the sum of boundaries settled; lighter tint denotes boundaries remaining by continent. Altogether 161 countries across six continents.

Figure 2.5   *Maritime boundaries settled/remaining per continent*

Unsurprisingly, large countries with more access to maritime space have a larger number of maritime boundaries. Australia, China, Canada, Norway and Russia have long coasts, resulting in multiple neighbours and, in turn, multiple maritime boundaries. Also, areas like the Mediterranean and the Caribbean, where numerous small states are clustered together, have a large number of maritime boundaries. Moreover, countries with overseas colonies or dependencies—such as France, the United Kingdom, Spain and the United States—have multiple maritime boundaries, both settled as well as unsettled.

With 180 boundaries at sea still not agreed on, it is likely that maritime boundary disputes will linger on the national and international agendas in decades to come. Some will remain dormant and rather insignificant, whereas others might flare up due to climatic, economic and/or political changes. Therefore, it is crucial to examine some specific trends that impact boundary-making at sea, to which I explore in the following chapters.

NOTES

1.   Spruyt (1994) takes Tilly's argument and qualifies it. He holds it was not war but rather the underlying capacities of states that determined their success and path towards the 'national state'. This capacity was determined by trade, which had come as an 'exogenous shock' between 1000 and 1400, in turn leading to various forms of class alliances that determined the abilities of states in an increasingly competitive environment (1400–1600). Both Tilly and Spruyt argue that by the Peace of Westphalia (1648), the sovereign state system had taken root and the 'national state' model had become the norm.

2. For a lengthy examination of the concept of territoriality and fixed territory, see, for example, Elden 2013; Agnew 1994; Storey 2012. Territoriality can be defined as the process whereby territory (here: the ocean) is claimed by individuals or groups. 'Territoriality can be seen as the spatial expression of power and the processes of control and contestation over portions of geographic space are central concerns of political geography' (Storey 2012, 8). Studies of territory and territoriality are primarily concerned with land and the human need/desire to inhabit and control land. However, the idea of 'socialised territoriality' is also relevant for discussions of the maritime domain because it enables the role of territory to be conceived of more broadly. Sack (1986, 219) sees territoriality as a 'device to create and maintain much of the geographic context through which we experience the world and give it meaning'. In turn, once 'territories have been produced, they become spatial containers within which people are socialized' (Storey 2012, 20; see also Paasi and Prokkola 2008).

3. Because the European colonial states had largely destroyed indigenous political structures, setting up only rudimentary 'Western' state-structures and then leaving behind a limited administrative infrastructure, the newly independent African states were not able to effectively control their territory and the attributes generally associated with statehood (Jackson 1987, 526–28). They obtained 'judicial statehood', which implies negative sovereignty, instead of empirical statehood (positive sovereignty) previously associated with states and their territory. The leaders of the newly independent countries in the 1950s and 1960s chose to (or had to) keep their somewhat artificial boundaries because they lacked the means to consolidate state territory properly. The point here is that the timing of independence from European colonizers, the access to foreign capital and the interference by European/Western powers are all important dimensions related to state formation. These factors had not been present in the European experience half a century earlier.

4. In Asia, a study by Hui (2014) contrasts the European experience of interstate war with China's intrastate war and how this led to a different outcome in terms of state capacity. Scott (2009) argues that population density was a strong determinant for the trajectory in South-east Asia because it was not territory itself, but the size of the population that was deemed important for state formation.

5. Zacher adapts this from Finnemore and Sikkink (1998).

6. One nautical mile (n.m.) is 1852 metres or approximately 1.15 miles, and this has become the standard unit of measurement for both marine and air navigation, as well as zones at sea.

7. LOS Article 74 concerning the EEZ has wording identical to that of the Continental Shelf, Art. 83.

8. 'The "Chancellor's foot" has ... become proverbial shorthand for the argument that equity is an unjustified and unfortunate interference in the regular course of the rule of law' (Powell 1993, 7).

9. See, for example, Árnadóttir 2016.

3.   Explaining maritime boundary dispute settlement

We have seen how the contemporary concept of a boundary at sea came to the fore in the international system, that is, the rules that govern 'who owns what' at sea. By linking it to state formation itself and the idea of borders on land as a method for delineating territory, I have displayed how boundaries at sea are related to, yet different from, similar concepts on land. The legal regime that took shape over the course of the past century came to define precisely what rights states could claim at sea, so processes were developed for settling the disputes that inevitably arose. Because these hinge on political considerations and context beyond purely legal principles and/or geography, we need to employ a wider lens in exploring maritime boundaries.

This chapter outlines how political scientists and lawyers have approached the topics of maritime space and settling maritime boundary disputes. The literature on territorial disputes on land is broad and well established, arguably constituting its own subfield of conflict studies and/or IR. It covers various ways of conceptualising foreign policy and the international system (across theoretical paradigms) while also grappling with the continuing challenge of how to bridge the international and domestic levels of analysis. Concerning the maritime domain, however, the gap between IR and international law is wide: the legal literature often does not engage with the political context, and the political science literature neither examines the maritime domain spatially nor explores its connection to international law.

3.1   POLITICAL SCIENCE AND THE INTERNATIONAL SYSTEM

Much ink has been spilled on conceptualising the seas as a domain for power projection and sovereignty. As states operate in a system of anarchy, typical 'realist' thinking would argue that maritime boundary disputes are inherently left to the states themselves to solve and that, in reality, international authorities and/or courts do not have significant roles in mitigating or handling such disputes (Waltz 1979, 204–9). State survival takes primacy, so strategic/security considerations concerning the given territory become paramount (Gilpin 2001, 19). Further, strategic rationales concerning the security value of a territory tend to be a key driver of conflicts over territory. Carter (2010) argues that particularly weak states often use territorial consolidation for strategic advantages to improve their relative position in an inferior power balance. Thus, when the territory in question gives the state a strategic security advantage, the likelihood of conflict will be high (Oosterveld, De Spiegeleire, and Sweijs 2015, 21).

Several other studies have noted the potential for maritime conflict over resources and/or the symbolic value of spatial domains.[1] Hensel et al. (2008) examine how certain issues have a higher priority than others for states in the international environment. Employing a division between tangible and intangible values, they distinguish between salient tangible (economic, strategic) and salient intangible (prestige, identity) issues. Basically, they argue that issues concerning rivers and the maritime domain are tangible but lack the intangible dimension that territorial (terrestrial) disputes possess. In turn, this makes maritime issues less likely to reach the top of the political agenda and lead to conflict (Hensel et al. 2008, 121, 138–40).

Still, disputes at sea may also be used to showcase structural power balancing and constraints (Tunsjø 2018). Nyman (2013; 2015) explores what drives conflict over maritime issues more generally, finding that states do indeed engage in conflict when resources are involved. Like Kleinsteiber (2013), she focuses on both the resource dimension and domestic and national interests in maritime disputes, advancing our understanding beyond merely dismissing the maritime as limited in salience.

Osgood also notes that the ocean has always held a crucial role for military power projection but that it is nonmilitary utilisation of the ocean that has led to its primacy in national and international affairs (1976, 10–12). Building on such accounts of the ocean, Till (2004) describes oceans as having a growing relevance as a dominion of power through two complementary variants of sovereignty: instrumental and expressive. Maritime sovereignty may be integral to the state's survival (the former), or upholding sovereignty over a maritime domain may become an expression of 'national pride and effectiveness' (the latter) (Till 2004, 289).[2]

These works, however, as well as more recent conceptions of the ocean as a power base or strategic domain,[3] are limited to two interrelated relationships. First, they are concerned with how oceans influence the security outlook of states. Second, they are interested in how states utilise the ocean for the same purposes: boosting their geopolitical standing vis-à-vis other states. In these accounts, the ocean is arguably no different from other spatial domains—it is depicted as interesting because of its special characteristics.

In this book, however, the focus is not on the impact of maritime space on the state (or its security) but instead on the relationship between states and the ocean as a space that relates to, but is distinct from, land, when states engage in interactions among themselves over rights and access to maritime space. In other words, it is not how geopolitics are impacted by the sea, but rather the geopolitics taking place at sea over issues at sea that matter. The emphasis is on how states approach this space, both legally and politically, vis-à-vis each other and how this interaction is changing. Studying how states deal with disputes over where to delineate rights offers insights into the larger relationship between society and ocean as a spatial domain, going beyond historic arguments or the importance of the ocean for state power.

A variant of this comes from Steinberg's (2001) The Social Construction of the Ocean. He examines how the idea of maritime space has changed throughout history. There are many ways of thinking about the ocean: as a territorialised extension of land; as a domain where only limited control can be exercised; and as a great void (Steinberg 2001, 18–25). In turn, he argues, states have desired to keep the oceans free of conflict. Baker (2013, ii) supports this, finding that states have become behaviourally conditioned by an international norm against the 'forceful acquisition of maritime spaces and resources of other states'.

3.2   INTERNATIONAL INSTITUTIONS AND THE LAW OF THE SEA

In contrast or in addition to traditional system approaches to IR, legal or 'institutionalist' approaches place emphasis on group interests and the constraining and/or enabling traits of the institutional structures (formal or informal) that states have established to jointly manage problems in the international sphere (Young 1986; 2011; 2021; Tarrow 2001; Keohane and Nye 1977). Institutionalist approaches see international regimes as helping to stabilise the contractual environment while enabling states to implement agreements (Levy, Keohane, and Haas 1993; Young 2012; 1989; Levy, Young, and Zürn 1995). International law—or specific aspects of it—are seen as international regimes, with the ability to constrain and alter the behaviour or otherwise independent states (Finnemore and Toope 2001; Reus-Smit 2004). The central points here are the conceptualisation of interests driven by factors beyond (or despite) security, as well as the influence of international regimes and institutions on the states involved.

For example, Allee and Huth (2006) note that states have complied with virtually every ruling made by the ICJ and its predecessor, the Permanent Court of International Justice. They conclude that leaders are more likely to make use of international legal institutions when facing strong domestic opposition to bilateral concessions in a settlement: the legal bodies can provide 'political cover' (Allee and Huth 2006, 300–301). Moreover, testing their hypotheses on a large number of cases, Huth, Croco and Appel (2011) have shown how leaders with strong legal claims are more likely to push for negotiations to reach a settlement and more likely to avoid the use of force if they are dissatisfied with negotiation outcomes. Asymmetry between the legal claims of states are also more likely to lead to a settlement than are situations where the claims are balanced. Further—and supporting the 'domestic cover' theorem—leaders with weak legal claims are more likely to seek settlement through a formal dispute resolution. In sum, these authors find strong support for the argument that 'international law matters', in contrast to the traditional realist theorem that state actions are guided by power and interests alone (Huth, Croco, and Appel 2011, 433).

Maritime disputes are no exception because formalised cooperative structures might help the settlement process along, and international legal scholars have thoroughly grappled with the development of the LOS.[4] For example, a coerced outcome to a maritime boundary dispute would be contrary to Article 2(4) of the UN Charter and, therefore, would lack international legitimacy. The point here is that the use of sheer power in determining the limits of state jurisdiction has been ruled out in the post-Second World War order.

It is within this environment that the LOS has developed and serves as framework—or international regime—with the approaches and tools needed for settling boundary disputes (Keohane and Nye 1977, chap. 4). Thus, the LOS is part of this larger legal–political reality. As Keohane and Nye (1977, 56) put it, '[T]here is very little direct functional relationship between fishing rights of coastal and distant-water states and rules for access to deep-water minerals on the seabed; yet in conference diplomacy they were increasingly linked together as oceans policy issues'. The LOS regime plays a vital role in managing maritime disputes by providing the mechanisms and procedures for states to manage and conclude agreements.[5]

Still, legal scholars recognise the political perils of maritime space. As argued in *Maritime Boundary* by Jagota (1985, 4), 'Maritime boundary, like territorial or land boundary, is a politically sensitive subject, because it affects the coastal State's jurisdiction concerning the fishery, petroleum and other resources of the sea as well as concerning the other uses of the sea'. Weil (1989, 30–31) further argues in *The Law of Maritime Delimitation – Reflections* that 'maritime boundaries, like land boundaries, are the fruit of the will of States or the decision of the international judge, and neither governments nor judges limit themselves simply to scientific fact'.

A few international lawyers have attempted to outline the relevant political dimensions. In his 'International Maritime Boundaries: Political, Strategic, and Historical Considerations', Oxman (1995) lays out in a fairly straightforward way the various political dimensions that might impact boundary-making at sea. He explains the importance of having a specific economic interest in a given area and subsequent moves by a given state to 'stimulate uses' (Oxman

1995, 247). A similar approach is found in Johnston's *The Theory and History of Ocean Boundary-Making* (1988).[6] Written as a comprehensive overview of all the factors of relevance to maritime boundaries and boundary dispute settlement, it tackles the topics here but discussed from the political and legal angles.[7]

3.3   LOOKING INSIDE THE BLACK BOX

Albeit crucial to understanding geopolitics at sea, the legalistic approaches described are often overly concerned with the process itself, ignoring many of the factors within the international system that are pertinent to the outcome (Hafner-Burton, Victor, and Lupu 2012). If we truly want to comprehend how states behave at sea, an additional approach is to take national preference formation seriously (e.g., Moravcsik 1997). As several scholars have shown, the interaction between the domestic and international levels of analysis runs both ways. Gourevitch (1978) shows that the international system impacts domestic set-ups and the development of states. Putnam (1988) theorises a two-level game between the domestic and international level in terms of international negotiations.

Globalisation has also eroded the distinction between the domestic and international (Milner 1998). Keohane and Nye (2012, org. 1977) note that the growing numbers of transnational actors operating beyond the realms of the state have led to challenges for state autonomy in the international realm. As shown by Keck and Sikkink (1998) and Tarrow (2001), transnational advocacy groups and international networks have become increasingly important for any issue on the domestic level. Similarly, many multinational companies have operating budgets larger than those of some small states, wielding considerably greater leverage in the international system.

From a foreign policy perspective, state leaders make the final decisions regarding when to push for and accept boundary settlements. Thus, it is reasonable to assume that why a settlement occurs relates to a change in the utility of options available to the domestic leadership, rendering an agreement more advantageous than the status quo (no settlement).

Scholars have often turned to specific case studies of single maritime boundary disputes, countries or regions/domains to investigate the politics and interests of states at sea. Ásgeirsdóttir (2016, 195) proves this by displaying the importance of domestic public opinion in settling boundary disputes when examining US maritime boundaries specifically. Domestic structures have an impact on states' preference formation and, thus, on the outcomes of negotiation. Oude Elferink (2013) shows how the outcomes of the North Sea boundary disputes in the 1960s were not just a matter of international law and the ICJ ruling: it was also heavily influenced by political aspirations and interests in the three countries involved.

In his study of the relationship between Canada and the USA, McDorman (2009, 3) finds that '[o]ne of the reasons for the pragmatism of the two States in avoiding legal and political confrontation is the fear of resolving a dispute'. In other words, the two states are concerned with negative domestic public responses to a possible resolution and, thus, would prefer to keep the dispute pragmatically managed, at least for the time being. Further, Ø. Jensen (2014a) finds that Norway has employed a particularly proactive approach to the LOS regime, attempting to settle disputes to avoid protracted litigation or negotiation (Ø. Jensen 2014a, 50–55). In other words, political leaders have *agency* when it comes to settling their maritime disputes. This is confirmed across several single-boundary case studies[8] or those looking at regional relations and disputes.[9]

3.4   THE PROCESS OF SETTLING A DISPUTE AT SEA

One the one hand we have the overarching explanations for why states agree when they do. On the other hand, is how they get to that point? First and foremost, for a maritime boundary to be settled, it needs to be on the agenda and, thus, in the 'policy process' of the relevant states. This is not always the case, though. Initially, when states enlarged their maritime zones after the Second World War, disputes emerged. LOS Articles 74 and 83 regarding the EEZ and the continental shelf specify that states 'shall make every effort' to settle outstanding disputes (United Nations 1982). However, as history has shown, when states are unable to settle disputes immediately, they often became entrenched or simply neglected (Wiegand 2011a). Now and again, such disputes resurface on the policy agenda and may be settled or potentially fuel conflict.

Why do some disputes reach the political agenda, whereas others are kept out of the (political) limelight? Within the field of IR studies, 'agenda setting' often refers to the role of the media in defining the agenda of states and their elected politicians (Manheim and Albritton 1984; Scheufele and Tewksbury 2007; Robinson 1999; 2008; Livingston 1992). However, we can progress beyond this focus by examining literature within the field of public policy. Initially, the political agendas of states were thought to be a product of their economic environment and level of industrialisation. The policies of industrialised countries were thought to converge within the same policy mix (Howlett, Ramesh, and Perl 2009, 94).

As scholars came to recognise the importance of ideas and beliefs, new theories of agenda setting emerged. Goldstein and Keohane (1993) show how worldviews, principled beliefs and causal ideas are all relevant to policy-making. The sentiments and beliefs (or identities) held within a community may be shaped by and in turn shape the agenda of that community (Campbell 1998). Similarly, the political discourse used within a community is central to understanding where the 'policy problem' has arisen from (Foucault 1972). Not all actors within a community hold the same beliefs to the same degree, so agenda setting often entails a clash between different beliefs and convictions.

In response to the focus on structural and economic factors, scholars in the 1970s developed the 'funnel of causality'—a concept including all the variables of the socio-economic environment, distribution of power, prevailing ideas and ideologies, the institutional framework of government and the decision-making processes (King 1973; Howlett, Ramesh, and Perl 2009, 99). Although this concept has been criticised for including too much, thereby becoming too context specific, the greatest strength of the 'funnel' is its causal variance and capacity to explore the relationships between different sets of variables (Mazmanian and Sabatier 1980).[10]

Further, Kingdon (1984) identifies what he terms 'policy entrepreneurs' and 'policy windows'. Identifying the former is essential for understanding an issue and where it comes from. The latter concerns whether or not entrepreneurs are successful in their efforts, and the ability to explain why some issues (and not others) reach the agenda (Wood and Peake 1998).[11] Some windows are fairly predictable (e.g., legislative sessions); others are random (e.g., catastrophes). In any case, open windows are 'small and scarce' (Kingdon 1984, 213).

Are foreign policy issues different from those of domestic public policy? This question speaks to the core of IR and the question of where state policies and interests derive from. I limit this debate by stating that in discussing maritime boundary disputes, I recognise that agenda setting power and influence may derive from both levels. In other words, foreign policy outcomes are a result of the relevant factors at both the domestic and international levels.

As to territorial disputes specifically, the question is to what extent these reach the political agenda at all. Wiegand (2011a) argues that one problem is that some territorial disputes fail to reach the political agenda because they lack salience, so no real attempts to initiate settlement occur. Disputes may also become linked to other issues, making them more difficult to solve. Therefore, we need to distinguish not only among the different sources of policy initiatives and their 'windows', but also among the issues involved. 'Because diverse issues are valued differently by states, this also means that the degree of salience varies depending on the issue and whether it has low or high salience for states' (Wiegand 2011a, 46).

Where do maritime boundary issues fall within these categories? What is the 'salience' of maritime boundary disputes in general, and why and when do they find their way onto the political agenda? An inherent recognition is that any given state's resources to spend on bilateral negotiations on maritime boundaries are bound to be limited in terms of capital, but even more so in terms of highly qualified senior diplomats and legal experts with the skills and capacity for these processes. Therefore, the fundamental premise of the current study is that states will engage in maritime boundary dispute negotiations only when they perceive there will be significant gains.

Accepting the wider 'funnel of causality' approach, we must also recognise that there may be more than one reason why a dispute receives political attention. It is essential to comprehend the broader context in which a given dispute arose on the political agenda and the processes that determine the outcomes being studied. This brings us to the crucial distinction between the impetus of negotiations in the first place and what drives or hinders the settlement of a boundary dispute. We can nuance the somewhat simplified dichotomy between settled/not settled regarding maritime boundary disputes. Figure 3.1 envisions a range of options as ordinal categories:

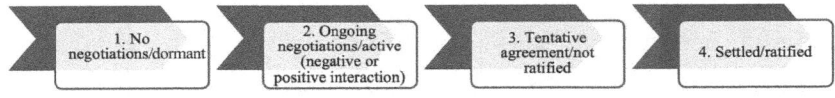

Figure 3.1   The process of settling a maritime boundary

Conceiving of outcomes as a process enables us to identify the impetus for negotiations and distinguish this from what affects states' willingness to compromise and concede in these negotiations, along with being able to distinguish those factors that might drive negotiations forward, step by step, from those that might obstruct the process when moving along the same path. Let us now examine how and why some specific countries have settled their maritime disputes or have not been able to do so.

NOTES

1. See, for example, Wiegand 2011b, 2011a; Kaplan 2011; Carter 2010; Diehl et al. 2006.
2. Mearsheimer (2001), however, argues that water actually serves to impede power relations: power dynamics become reduced over bodies of water.
3. Such as Kraska's *Maritime Power and the Law of the Sea*
4. See Oxman 1995; Ravin 2005; Harrison 2011; Weil 1989; D. M. Johnston and Saunders 1988; D. M. Johnston 1988; Vidas 2018; Oude Elferink, Henriksen,

and Busch 2018b; 2018a; Busch 2018; Irwin 1980; Brown 1981; Klein 2006; Oude Elferink 2015; McDorman 2002; Ø. Jensen 2014b; Lalonde 2010; Oude Elferink and Rothwell 2004; Nasu and Rothwell 2014; Cottier 2015; Blake 1994.

5. For more on this, see, for example, St-Louis 2014; Nelson 1990; Gray 1997; Kwiatkowska 2004; Ravin 2005; Vidas 2018; Busch 2018.

6. It can be discussed whether Johnston's work should be placed in the legal or political section because Johnston was a political scientist who ventured into the realms of international law and LOS in his work at the Schulich School of Law at Dalhousie University.

7. Several works provide *guides* to the practice of agreeing on maritime boundaries, such as *A Practitioner's Guide to Maritime Boundary Delimitation* (Fietta and Cleverly 2016) or *Maritime Border Diplomacy* (Nordquist and Moore 2012).

8. Baker and Byers (2012) hold that the increasing attention given to the Arctic region and potential of marine resource extraction there has raised the Beaufort Sea dispute on the agenda; in a study of a maritime dispute between Bangladesh and Myanmar, Bissinger (2010) argues that the economic potential from gas prospects drove state interests; Schopmans (2018) examines cooperation on marine resources concerning the Falkland Islands; Siousiouras and Chrysochou (2014) study the complexities of the Aegean Sea disputes between Greece and Turkey; Wiegand (2012) explains a dispute concerning the islands between Bahrain and Qatar; Eiran (2017) compares the differing conceptions of sovereignty relating to land and sea in the ongoing conflict between Israel and Lebanon; Song (2015) examines how fishing vessels are used in the ongoing dispute between North and South Korea; Vidas (2009) shows how Croatia, Italy and Slovenia managed to resolve a long-standing dispute in the Adriatic Sea; Supancana (2015) explores the maritime boundary dispute between Indonesia and Malaysia, focusing on an area rich in resources and while providing possible resolution mechanisms; Thao and Amer (2007) examine Vietnam's multiple maritime boundary issues, many of which have remained unresolved; Qiu and Gullett (2017) perform a quantitative analysis aimed at improving on the ITLOS verdict concerning Bangladesh and Myanmar; Hasan and Jian (2019) explore the long-standing boundary dispute in the Bay of Bengal between Bangladesh, India and Myanmar; and Walker (2015) surveys the numerous African boundary disputes, making the overarching claim that settling them would be good for the individual countries. Similar ideas are taken up in greater depth by Okonkwo (2017, 76).

9. The South China Sea in particular has attracted many case studies, ranging from elaborations on the complexities of the disputes (Berakit 2011; Ong 2015; Hong 2010; Davenport 2013), explanations of Chinese perspectives and positions (Zaibang 2012; Jianfei 2012; Junhong 2012) and proposed solutions to future challenges in the region (Beckman et al. 2013; Beckman and Schofield 2009; Kleine-Ahlbrandt 2012; Cheng and Paladini 2014; Mishra 2017; Schofield, Sumaila, and Cheung 2016; L. C. Jensen 2014). The East China Sea dispute over the Senkaku/Diaoyu Islands, as well as the Japan–Russia dispute over the Kuril Islands, are also often cited to underscore the relevance of the topic of territorial disputes (Smith 2012; Hirano 2014; Kaczynski 2007; Wiegand 2011a; Cui 2014). Similarly, the attention given to the Arctic at the start of the twentieth century has led to an examination of the regional dynamics regarding maritime disputes (e.g., Byers 2013; 2009; Hoel 2014; Churchill 2001; Østhagen and Schofield 2021; Schofield and Østhagen 2020; Lalonde 2010). For example,

Moe, Fjærtoft and Øverland, looking at the 2010 Barents Sea delimitation agreement between Norway and Russia, note various relevant explanations for this breakthrough, such as Russian efforts to tidy up border disputes, Russia's desire to be perceived as a constructive international actor and its desire to reaffirm the LOS regime (Moe, Fjærtoft, and Øverland 2011). These findings were replicated in a similar article by Orttung and Wenger (2016). A related study by some of the same authors has examined exactly how the two states developed a joint regime for oil and gas in the formerly disputed area (Fjærtoft et al. 2018).

10.   Beyond these initial conceptions, various modes and cycles have been identified by public policy and agenda-setting scholars. Downs (1972) argues that issues that arise on the government's agenda behave similarly to the media's news cycle: they fade as the issue at hand becomes more complex than initially thought when it was first sensationalised. Later, this proposition was further specified into cycles initiated by external events (such as war), here termed a 'crisis cycle', and cycles initiated by the political leadership, called 'political cycles' (Howlett, Ramesh, and Perl 2009, 101).

11.   Cobb, Ross and Ross identify the different models of agenda-setting. The *outside initiation model* is found predominantly within a liberal pluralist society when an issue is brought to the agenda by an interest group external to the government and the policy processes. The *mobilisation model*—associated with totalitarian regimes—concerns issues placed on the agenda by decisions of the government and/or leadership without input from the public (in this phase of the policy cycle). Finally, the *inside initiation model* takes into consideration influential groups with access to the government that initiate policy, where the process unfolds in mutual consultation but without engagement from the broader public (Cobb and Elder 1972; Cobb, Ross, and Ross 1976, 134–36). The division between liberal and totalitarian regimes has been criticised, but the initial categorisation and conceptualisation of actor engagement in agenda-setting models remain valid (Howlett, Ramesh, and Perl 2009, 103).

4.   Australia – the oceanic continent

In Australia's early colonial history, the maritime domain served as a buffer against potential security threats originating in the near Pacific.[1] 'Australia is an island continent (see Figure 4.1), the only nation-state to have a major land-mass to itself' (McCreery and McKenzie 2013, 560). From the 1950s onwards, new and industrialised ways of fishing, as well as the potential for offshore hydrocarbon resources and minerals, led states, including Australia, to turn their focus seawards (Lowe 2013). Australia borders or has access to several oceans and seas: the Tasman Sea to the southeast, the Java, Timor, Arafura and Solomon Seas to the north, the Coral Sea to the northeast and, more generally, the Indian Ocean to the west, the Pacific Ocean to the east and the Southern Ocean to the south.

Source: https://commons.wikimedia.org/wiki/File:Australia_states_and_territories_labelled.svg.

Figure 4.1   Map of Australia's territory, including external territories

The initial drive for marine resources in Australian waters arose because of the abundance of whales. As explained by Gaynor (2013, 275): 'The industrial revolution in Britain had given rise to an insatiable demand for whale oil: the spinning frames of northern England were lubricated with it, and lamps burning it illuminated factories long into the night.' As whale stocks in the northern waters became depleted, the industry shifted its focus south towards species that migrate to and from Antarctica. However, having expanded rapidly, the whaling industry was in decline by the 1860s due to overharvesting. It is estimated that the whale population has still not regained pre-eighteenth-century levels (Gaynor 2013, 277). The regional seal population faced a similar fate. Consequently, industrial ventures turned their focus inland towards mining (predominantly of gold) and pastoral activities (Gaynor 2013, 280–85).

It was not until 1990 that Australia extended its territorial sea to 12 n.m., announcing its intention to establish an EEZ just one year later (Burmester 1995, 52). Australia implemented an EEZ in 1994, having preferred until then to refer to it as the 'Australian Fishing Zone' (Forbes 1995, 101). This fishing zone, based on the same principles as an EEZ, had been established in 1979 (Kaye 2001, 11). Schofield (2008, 4–5) describes the Australian approach as 'conservative, cautious and orthodox, largely because of the relatively slow pace at which Australia has adopted extended claims to maritime jurisdiction'. This is also related to fear of opposition from third parties to Australia's long baselines and historic claims, as well as the relative remoteness of most of its maritime space (Schofield 2008).

Australia has entered into maritime boundary treaties with Indonesia, Timor-Leste, Papua New Guinea, Solomon Islands, New Zealand and France, in effect settling all its maritime boundaries with its neighbours,[2] with the latest settlement coming as recently as 2018 with Timor-Leste. However, this overall achievement did not come without costs and can also be seen as the result of continuous efforts at pursuing settled maritime zones starting in the early 1970s.

4.1   INDONESIA

Starting in 1971, Indonesia and Australia agreed on a seabed boundary, delineating the boundary between Papua New Guinea (then administered by Australia), Australia and Indonesia itself. Equidistance was the principle method utilised for boundary drawing, as referred to in the treaty itself (Australia–Indonesia 1971; Kaye 2001, 46). The agreement in 1971 specifies a provisional joint regime for equitable exploitation of straddling seabed resources, when needed.

The following year, in 1972, the two countries agreed on an extension of this continental shelf boundary, stretching through the Timor Sea from the agreed point in the Arafura Sea. This included a delineation between the various Indonesian islands in the northern part of the Timor Sea, including the island of Timor itself. The agreement, which was negotiated in 1972, was influenced by the outcome in the 1969 North Sea cases, which can be seen by the fact that Australia leaned on the ICJ's verdict to argue for a larger share of the maritime space up to the point where the shelf drops into the Timor Trough at 3000 metres depth (Prescott 1993d, 309), not the median line as argued for by Indonesia.

Further, Australia (on behalf of Papua New Guinea) and Indonesia concluded an agreement on a small remaining gap between land and a point where the 1971 seabed boundary started because there had been some uncertainty as to the point on land from which to delineate the territorial waters. By 1973, this had been settled (Australia–Indonesia 1973). These agreements did, however, cause negative reactions in Indonesia. Australia was perceived as gaining the most through the negotiations. As the former Indonesian foreign minister put it, Indonesia had been 'taken to the cleaners' (Kaye 2001, 21).

Although seabed issues had been settled in the agreements discussed above, fisheries relations remained a separate issue. In 1979, Australia declared a 200 n.m. fisheries zone that overlapped with the zones of some of its neighbours. Australia also claimed the full zonal effect of the uninhabited Ashmore and Cartier Islands located close to Timor, to which Indonesia objected. Already, in 1974, the two countries had signed a Memorandum of Understanding (MoU) concerning traditional fisheries in the maritime areas around the Ashmore, Cartier and Browse Islands (Australia–Indonesia 1974). This agreement aimed at allowing 'traditional' Indonesian fishers to fish without motorboats or electric fishing equipment and to make use of the islands for water supplies.[3]

However, during the 1970s and early 1980s, several negative incidents involving overfishing and a breach of the MoU damaged fisheries relations in the area (Prescott 1993a, 1234), in turn prompting further negotiations on a final maritime boundary. In 1980/1981, Australia and Indonesia agreed in two rounds of negotiations on a provisional fisheries surveillance and enforcement line. This agreement was signed on 29 October 1981. In sum, this agreement concluded a highly complex boundary-making exercise in the Timor and Arafura Seas between Indonesia and Australia (Timor-Leste was later added to the mix).

However, the problems around traditional fisheries in the boundary area did not disappear. At times, forceful behaviour – with the Australian authorities burning Indonesian boats – caused outcry and media headlines (Stacey 2007, 2). The final agreement was signed on 14 March 1997, concluding almost 30 years of negotiations between the two neighbours (Australia–Indonesia 1997). As of the time of writing this book, the agreement has still not entered into force because both countries have yet to ratify it, even though both countries act according to it.

4.2   PAPUA NEW GUINEA

In 1971, before Papua New Guinea became independent, Indonesia and Australia agreed on a maritime boundary deriving from the border on land that delineated the seabed between Indonesia and Papua New Guinea. When Papua New Guinea gained independence from Australia in 1975, an agreement on a maritime boundary between the two countries was needed. Negotiations had been initiated in 1972 and started in 1973 with independence being imminent (Kaye 2001, 102; Forbes 1995, 120). Naturally enough, the initial impetus for boundary negotiations came from Papua New Guinea's aspirations of achieving independence from Australia. The treaty was signed on 18 December 1978 but entered into force only on 15 February 1985, after both the Australian Parliament and the Parliament of the Australian state Queensland, as well as the Parliament in Papua New Guinea, had examined the agreement and deliberated on it (Australia–Papua New Guinea 1978).

The treaty took over six years to negotiate and is highly complex – or as Oxman (1995, 249) terms it, 'unusually sophisticated' – with several distinct elements. This package deal involved drawing four different boundaries at various locations (Park 1993b, 931): a seabed boundary; a fisheries jurisdiction boundary; a combination of the two; and a protected zone (Willheim 1989; Kaye 1994). In particular, opposition from Queensland, which was unwilling to abandon its inhabitants on the small islands close to Papua New Guinea, was seen as central (Burmester 1982, 327).

Similarly, yielding sovereignty over territory was seen as constitutionally difficult in Australia. Only after national elections in both countries in 1977 – and with Papua New Guinea expanding its territorial waters to 12 n.m. and implementing a 200 n.m. EEZ in 1977 – did the two countries push the issue forward. In advance of the final settlement in December 1978, the decision by Australia to affirm sovereignty over the above-mentioned islands was made public, without giving rise to strong counter-reactions (Burmester 1982, 327). Forbes (1995, 122) holds that this illustrates how 'by mutual agreement, countries are able to reconcile their differences when there are benefits to be gained in determining the most appropriate means of sharing the resources of the seabed adjacent to their respective coastlines'.

4.3   FRANCE

What appears to be one of the least challenging maritime boundaries for Australia to settle came with France in 1982 when the two countries agreed on two boundaries simultaneously. This 'Agreement on Maritime Delimitation' was signed on 4 January 1982 and entered into force on 9 January 1983 (Australia–France 1982). In the context of the third round of LOS negotiations and the realisation that 200 n.m. was becoming a widely accepted limit, both France and Australia showed interest in finalising maritime boundaries with their neighbours (Park 1993a, 906). Park further notes that there were no economic aspects to the negotiations, which was why the parties managed to conclude the agreement in the course of only three days.

The second boundary established with this treaty was between the Heard and McDonald Islands (Australia) and Kerguelen Island (France) in the southern Indian Ocean. These islands are located in relatively extreme environments, with little or no permanent human settlements (Gullett and Schofield 2007, 547). The boundary was drawn between the Australian fishing zone and the French EEZ, as well as delineating the continental shelf between the two parties. There is no provision for potential mineral or hydrocarbon fields that might straddle the agreed boundary, which adds weight to the argument that economic considerations – as in the case with New Caledonia – played a limited role in shaping the outcome. Also, given these islands' proximity to Antarctica, their use for research purposes was deemed more important than for economic ventures (Prescott 1993c, 1186).

However, after the boundary was settled, fisheries have become more of an issue regarding Heard and McDonald Islands and Kerguelen Island. In particular, Patagonian Toothfish became a commercially important species starting from the late 1990s onwards, spurring a number of illegal, unreported and unregulated (IUU) vessels in the area. Arrests were made by both France and Australia, eventually prompting two rather extensive and unique agreements in 2003 and 2007 between the two countries to allow the 'hot pursuit' of targeted vessels in each other's territorial waters (Gullett and Schofield 2007). In turn, it can be argued that the relatively uncomplicated establishment of a maritime boundary in 1982 paved the way for deeper cooperation between the two states as fisheries improved in the area and marine resources became more relevant.

4.4   SOLOMON ISLANDS

Australia and the Solomon Islands declared 200 n.m. fisheries zone and EEZs in 1979, one year after the latter had gained independence from the UK. Negotiations on a maritime boundary had already started in 1978, but it took a decade to reach an agreement because the other countries involved – Papua New Guinea to the north and France (New Caledonia) to the southeast – had an impact on where to draw the exact boundary. The boundary itself is not very long, only 150 n.m. (Park 1993c, 978).

The agreement was signed on 13 September 1988 and entered into force on 14 April 1989 (Australia–Solomon Islands 1988). It was based on equidistance from the basepoints of each state, serving as an all-purpose boundary delimitating both the EEZ and continental shelf (Kaye 2001, 142). Economic interests do not appear to have played a decisive role in the negotiations or the outcome because there was – and still is – only limited economic interest in this specific maritime area (Park 1993c).

4.5   NEW ZEALAND

New Zealand claimed an EEZ in 1977, while Australia waited until 1994 to do so. The distance between mainland Australia and mainland New Zealand is about 1200 n.m., here across the Tasman Sea. Because of Norfolk and Lord Howe Islands (Australia) and Three Kings Island (New Zealand) to the north and Macquarie Island (Australia) and Auckland and Campbell Islands (New Zealand) to the south, there arose a need to delimit EEZs and continental shelves. Australia and New Zealand had, however, informally agreed to an equidistant maritime boundary between their overlapping zones starting in the 1970s (Fyfe and French 2005, 3759).

Negotiations were motivated by submissions for an extended continental shelf to the UN CLCS, which was due within 10 years of the LOS entering into force in a country (Australia in 1994 and New Zealand in 1996). By 1999, as efforts were underway to finalise these submissions, the two countries officially started negotiations and agreed to complete their boundary negotiations by 2003. The treaty was signed on 25 July 2004, with entry into force on 25 January 2006. It sets out a relatively straightforward division of the EEZs, which are derived largely from the above-mentioned islands and not the mainland of the two countries. The islands on each side were comparable in size and distance from their respective mainland, which made it easy to apply the equidistance method. 'In fact, the agreement on the application of the median line for EEZ delimitation served to confirm a limit that "has been observed de facto by the two countries for more than two decades"' (Schofield 2008, 6).

4.6   TIMOR-LESTE

When Australia and Indonesia settled a continental shelf boundary and fisheries zone boundary in the early 1970s, a gap was created where the Portuguese colony of East Timor had claims to a maritime zone. In 1975, East Timor was annexed by Indonesia. The Indonesian and Australian governments signalled their intention to develop a zone of cooperation in the Timor Gap area after 1975, though it was not until 1989 that an actual agreement was announced. The treaty concerning the zone was signed on 11 December 1989, was ratified by both countries and entered into force on 9 February 1991 (Australia–Indonesia 1989; Forbes 1995). Oil and gas resources were the most obvious driver of this maritime boundary agreement (Prescott 1993b, 1249), as specifically mentioned in the preamble to the treaty (Australia–Indonesia 1989).

When the treaty was announced, it was criticised by Portugal, which argued for the right of self-determination by East Timor, as well as the inclusion of Portugal in the negotiations. Portugal referred to the fact that the UN General Assembly and the Security Council did not fully recognise Indonesian sovereignty in East Timor (Prescott 1993b, 1248). Simultaneously, domestic opposition in Australia, which was driven in part by Timorese refugees, stirred sentiments against the Indonesian governance of East Timor (Kaye 2001, 87–89).

By the 1990s, the independence movement in East Timor was gaining traction as the scope of Indonesian atrocities against the local population became clear (Dunn 2003). By 20 May 2002, a new government of East Timor was in place, and in September of that year, the country was renamed Timor-Leste. A new Timor Sea Treaty was negotiated and signed on 20 May 2002 (Australia–East Timor 2002). However, the new government in Timor-Leste rejected the treaty negotiated by the ENTAET. It argued that the 1972 agreement and the coordinates of the JPDA were inconsistent with international law (Schofield 2007, 198–99) and that the seabed boundary should have been based on equidistance instead of taking into consideration the nature of the continental slope (Prescott and Triggs 2005, 3809). Australia proposed a unitisation agreement with Timor-Leste – referred to as the Sunrise International Unitisation Agreement (Sunrise IUA) – which would 'validate existing production sharing contracts granted by Australia and enable exploitation to proceed' (Prescott and Triggs 2005, 3810). This was signed on 6 March 2003 but not ratified until 23 February 2007 because of controversies over boundaries and ownership. Also, in 2002, Australia withdrew disputes over maritime zones from the terms of its acceptance of compulsory jurisdiction in the ICJ and under the LOS, thereby ensuring that all maritime disputes would have to be negotiated directly between the parties.

Soon, however, another agreement was in place concerning revenue sharing. The treaty, in short form called the 'CMATS', was signed on 12 January 2006 and ratified on 23 February 2007 (Australia–Timor-Leste 2006). Still, the government in Timor-Leste was under domestic pressure to obtain concessions from Australia. It was criticised for the 'continuing perceptions that East Timor should have secured a significantly larger share of the seabed resources at stake' (Schofield 2007, 209). By early 2017, Timor-Leste had left the CMATS agreement. It had also taken the case concerning the maritime boundaries to arbitration at the PCA, only dropping the case after the Australian government agreed to renegotiate the boundaries (Doherty 2017). Australia agreed to negotiate under an LOS conciliation committee – the first of its kind to be used for such issues. After one year of negotiations, the parties signed a new agreement on 6 March 2018. This new treaty set out a completely new regime in the former Timor Gap, finally instilling an all-purpose maritime boundary for both the EEZ and continental shelf (Australia–Timor-Leste 2018).

4.7   THE OCEANIC CONTINENT

The simplest boundary agreements for Australia concerned small islands in the Pacific and the Indian oceans: with France over New Caledonia and Kerguelen, which are located on opposite sides of Australia (1982), and with the Solomon Islands (1988). These processes were fairly straightforward, where neither economic interests nor historical relations obstructed a mutually advantageous outcome based predominantly on equidistance delimitation in open ocean space. The same can be said for the 2004 boundary agreement with New Zealand, which also seems to indicate the relevance of close ties and cultural/historic bonds in facilitating processes (as specified in the treaty itself), prompting the two countries to agree on an extended continental shelf boundary in advance of submissions to the UN CLCS.

Otherwise, there have been boundary issues involving only three other countries – and Australia has spent considerable time reaching and developing agreements with these countries. The boundary with Papua New Guinea, a former colony of the UK and later Australia, proved the simplest to reach an agreement with but difficult to manage in practice. Signed in 1978 and in force by 1985, a complex regime was created by the division between a fisheries zone stretching to the shores of Papua New Guinea but wherein Papua New Guinean traditional fishermen have rights and the seabed boundary. Since then, challenges have arisen in upholding this regime, with adverse consequences being noted for local fishermen in Papua New Guinea.

The other complex regime involves Indonesia and Timor-Leste. With Indonesia, reaching an agreement was – as in the other cases – achieved rather quickly in the 1970s and later in 1981 after extended maritime zones were implemented. However, the separation between a seabed boundary and fisheries zone, as well as a zone where Indonesian fishers could continue 'traditional fishing', led to difficulties and negative publicity for the Royal Australian Navy (Akami and Milner 2013, 553).

These challenges link up with the Timor Gap created in the negotiations between Indonesia and Australia in the 1970s. The zone of cooperation in that area created in 1989 – after Indonesia had annexed the former Portuguese colony in 1975 – solved the problem temporarily but in a complex and – some would later argue – unfair manner. By 2002, the newly independent Timor-Leste wanted a better deal than the one negotiated with Indonesia.

Public resentment in Timor-Leste and in Australia led to a new treaty that was negotiated under a UN conciliation committee and finally agreed upon in 2018.

Resources seem to have been involved in several of the negotiations. This was the explicit – as stated in the treaties – rationale for the relatively simple boundary agreements with France, the Solomon Islands and New Zealand. The potential for hydrocarbons seems to have been a crucial factor in Australia's efforts to reach a compromise with Indonesia, Papua New Guinea and Timor-Leste, even if that meant giving up a larger share of the fisheries zone than might perhaps have been obtained through a principled stance.

There does not appear to have been a particularly strong military or security component in any of the boundary negotiations examined here. Although security matters have been discussed concerning the boundary with Papua New Guinea and, to some extent, with Indonesia, these issues focused more on safety and softer security issues, not traditional military alliances or outright conflict. None of the disputes has played a direct role in armed conflicts.

Thus, unlike the traditional theorems arguing for relative power considerations and strategic concerns, in the Australian case, both politicians and public opinion were concerned with being perceived as too ruthless in their negotiations with more vulnerable (at the time) negotiating partners. This speaks to the regional patterns of interaction that developed over a century from Australian independence to that of Papua New Guinea in 1975 and Timor-Leste in 1999. Further, some of these cases show that there need not be resources or an active conflict involved for states to find it mutually beneficial to settle their boundaries so that states can take a step in a larger strategy to complete their maritime maps and implement authority and control.

NOTES

1. See Lowe 2013.
2. This does not include the boundaries surrounding Antarctica, which are not dealt with here.
3. 'Traditional fishers' are defined as those who took fish using traditional fishing methods over a period spanning decades.

5. Canada – in the shadow of the hegemon

Canada has the world's longest coastline, facing three oceans: the Northeast Pacific, the Arctic and the Northwest Atlantic. The Canadian coastal areas in the Pacific and Atlantic hold – or have held – some of the richest fish resources in the world (Applebaum 2001). In 1969, the discovery of a major oil field at Prudhoe Bay, Alaska, also raised the prospect of oil and gas deposits in the Beaufort Sea, where the USA and Canada disagree on the location of the maritime boundary.

When international negotiations on the LOS started in 1973, Canada was interested in protecting its extensive fisheries to the east and west, especially in view of the importance of the Grand Banks fishing area off the coast of Newfoundland (Applebaum 2001). Between 1956 and 1977, Canada shifted from claiming the traditional 3 n.m. territorial sea to claiming a 12 n.m. territorial sea and exclusive jurisdiction over fisheries within 200 n.m. of its coast, as well as over mineral resources on its continental shelf. In 1977, the extension of the jurisdictions of fisheries by Canada gave rise to several boundary disputes in the Northeast Pacific, the Northwest Atlantic and the Arctic, predominantly with the USA but also with France (over Saint Pierre and Miquelon) and with Denmark (the waters between Greenland and Canada) (see Figure 5.1).

In 1977, Canada and the USA opened negotiations with the aim of resolving all four of their maritime boundary disputes; concerning Juan de Fuca, Dixon Entrance, Beaufort Sea and the Gulf of Maine. Prominent at the time was the dispute in the Gulf of Maine, which rests in the middle of a rich fishery that had previously been located in international waters (McRae 1989, 145–47). Canada began by expressing a willingness to grant concessions in the Beaufort Sea (Arctic) in return for US concessions seaward of Juan de Fuca Strait (Pacific) and, especially, in the Gulf of Maine (Atlantic) (Kirkey 1995, 55). It also sought a hydrocarbon-sharing regime for the Beaufort Sea so that oil and gas would not 'become a political or economic issue between the two countries because there would be joint access' and 'where the line was wouldn't make any difference' (Kirkey 1995, 55–56).

This attempt at a package deal failed because the USA insisted on dealing with each of the disputes independently and because Canada was concerned that in the absence of a package deal, a concession in one dispute could weaken

Source: https://en.wikipedia.org/wiki/File:CAN_orthographic.svg.

Figure 5.1   Map of Canada

its legal positions in the others. The US side was also worried about creating precedents regarding international law – not necessarily regarding disputes involving Canada but regarding disputes elsewhere.[1] In a standoff, the parties shifted their attention to just resolving the dispute in the Gulf of Maine, where immediate, competing economic interests made a solution imperative.

5.1   UNITED STATES

In 1977, Canada and the USA claimed fisheries zones out to 200 n.m. that overlapped on the eastern portion of Georges Bank (McDorman 2009, 135). Although Canada delimited its zone in the Gulf of Maine through a straight-forward application of the equidistance principle, the USA drew a modified equidistance line that took into account 'special circumstances', especially the shape of the seabed (McDorman 2009, 140–42). After three years of intense negotiations, Canadian and US negotiators signed two treaties in 1979 that were then sent to the US Senate for its 'advice and consent' to ratification. The East Coast Fisheries Agreement provided a complicated regime of transbound-ary fishing rights; however, it was never put to a vote due to opposition from the US fishing industry (McDorman 2009, 137). By contrast, the Agreement to Adjudicate the Maritime Boundary received the US Senate's advice and consent (US Senate 1981). In this second agreement, Canada and the USA agreed to submit the dispute to a chamber made up of five members of the ICJ (Valencia-Ospina 1996, 503), who were to delimit a single maritime boundary for both the continental shelf and EEZ.

In 1984, the chamber delimited a boundary out to 200 n.m. from the US coast that divided the disputed zone almost exactly in half (ICJ 1984). However, the end point of the adjudicated line was only 175.5 n.m. from the Canadian coast; as a result, 163 n.m.[2] of water column and seabed located within 200 n.m. of the Canadian coast were left unresolved. The USA has still not accepted Canada's jurisdiction to regulate fishing in that small area beyond the US 200 n.m. limit but south of the equidistance line (McDorman 2009, 176–78).

Another issue that remains is Machias Seal Island. The dispute over the island itself dates back to the 1783 Treaty of Paris, which assigned to the newly independent USA all islands within twenty leagues (60 n.m.) of its coast (Treaty of Paris 1783). In addition to the Treaty of Paris, the USA has based its position on the proximity of Machias Seal Island to the US mainland. In addition to the British land grant, Canada has based its position on the presence of a British (later Canadian) lighthouse on the island since 1832, which the USA did not protest until 1971.

Machias Seal Island and the surrounding seabed and waters have scant economic value. No oil or natural gas has been discovered in the area. Although the surrounding waters contain lobsters, which have been the subject of friction between Canadian and US fisherman, the potential fishery is not particularly large, and the two governments have exercised restraint, including adhering to a policy of flag-state enforcement (McDorman 2009, 193–94). Finally, any Canadian concession on Machias Seal Island would diminish the size of New Brunswick, thus bringing the interests (and perhaps constitutional rights) of that province into question. Similar considerations would seem to apply regarding the US state of Maine (Byers and Østhagen 2017).

Turning northwards, the Beaufort Sea is the shallow portion of the Arctic Ocean located between Alaska's and Canada's High Arctic islands, just north of the Mackenzie River delta. The dispute over the location of the boundary began in 1976 when the USA protested the line that Canada was using when issuing oil and gas concessions (McDorman 2009, 184). The dispute centres on the wording of a treaty concluded between Russia and Great Britain in 1825 (the USA took on Russia's treaty rights when it purchased Alaska in 1867; Canada acquired Britain's rights in 1880) (Great Britain–Russia 1825). This treaty set the eastern border of Alaska at the 'meridian line of the 141st degree, in its prolongation as far as the frozen ocean' (Great Britain–Russia 1825, para. 3). Canada claims that this treaty provision established both the land border and maritime boundary and that both must follow a straight northern line. In contrast, the USA holds that the delimitation applies only to land and that regular methods of maritime boundary delimitation apply beyond the coastline. In the case of the Beaufort Sea, the USA considers an equidistance line to be the legally and geographically appropriate approach (US Department of State 1995).

Each summer from 2008 through 2011, two icebreakers – one US and the other Canadian – worked together in the Beaufort Sea, gathering information about the shape of the ocean floor and the character and thickness of the seabed sediments (Boswell 2008; S. Griffiths 2010). This was a partnership borne of necessity because neither country had two icebreakers capable of the task and because both countries required a complete scientific picture of the seabed to determine the geographic extent of their sovereign rights to an extended continental shelf more than 200 n.m. from shore. This collaborative mapping beyond 200 n.m. may have also opened the door to a resolution of the boundary dispute by showing that the continental shelf in the Beaufort Sea might stretch 350 n.m. or even farther from the shore (Baker and Byers 2012).

Discussions were suspended at some point in 2011 after the two countries decided they would need more scientific information on the existence and location of hydrocarbon reserves before negotiating a boundary. Other factors probably included Cannon's departure from the foreign affairs portfolio, the mid-2011 fall in world oil prices and concerns about Canadian domestic law and public opinion (Byers and Østhagen 2017). World oil prices dropped sharply in 2014; and, in 2015, Shell shut down its activity north of Alaska, without having made noteworthy finds (Filippone 2015). Then, in December 2016, both the Canadian and US sides of the Beaufort Sea were put off limits for further oil and gas development as a result of a moratorium announced by the Obama administration and the new Trudeau government (Reuters 2016). As for fish, there is currently no commercial large-scale fishery in the Beaufort Sea, though indigenous peoples from both Canada and Alaska engage in some subsistence fishing.

One domestic impediment to resolving the boundary dispute could be the 1984 'Inuvialuit Final Agreement', a constitutionally recognised land-claims agreement in which the Canadian government and Inuvialuit used the 141° W meridian to define the western edge of the Inuvialuit settlement region (Inuvialuit Regional Corporation 1984). Finally, concerns about public opinion across the rest of Canada may have contributed to the suspension of discussions. During his nine years as prime minister (2006–2015), Stephen Harper had proclaimed himself a champion of Canadian Arctic sovereignty (Lackenbauer 2013; Byers and Webb 2013). Any concessions, especially to the USA, would have been treated harshly by the Canadian media and opposition parties.

Moving south-westwards, in 1903, the USA and Britain established an arbitration panel to delimit the border between the Alaska Panhandle and British Columbia (Canada–United States 1903). At the southern end of the panhandle, the panel drew a boundary down the middle of Portland Canal to just south of where it opens into the Dixon Entrance. However, the USA and Canada disagree on whether the specific points drawn by the panel concerns the maritime boundary.

In 1945, Canadian and US negotiators reached a tentative settlement of the Dixon Entrance dispute, whereby citizens of both countries would, outside the respective 3 n.m. territorial seas, have the right to fish and navigate on either side of an equidistance boundary. However, the Canadian national government withdrew, following objections from the provincial government of British Columbia (Bourne and McRae 1976, 215). Thus, the British Columbia government has involved itself in the Dixon Entrance dispute, blocking a tentative settlement in 1945 and issuing a position paper on the dispute in 1977 (British Columbia 1977).

Over the decades, both Canada and the USA have occasionally arrested each other's fishing boats in the Dixon Entrance. However, tensions over fisheries have subsided in recent decades, mainly for two reasons. First, in 1980, in an exchange of notes, the two countries agreed to observe flag-state enforcement (McDorman 2009, 285). Second, in 1985, they concluded the 'Pacific Salmon Treaty' and created the binational Pacific Salmon Commission to manage the fishery cooperatively along the entire coast (McRae 1989, 154–55). US Navy submarines regularly pass through the Dixon Entrance en route to an acoustic testing facility on Back Island, which is just north of Ketchikan, Alaska. In the early 1990s, Canada accorded navigational permission to the submarines, and the USA may have agreed to provide notice in advance of transits (McDorman 2009, 170–72). However, the USA has never accepted that Canadian permission is required.

Further south, the boundary between Canada and the USA within the Strait of Juan de Fuca was settled in 1846 (Great Britain–United States 1846), but the development of offshore rights in the mid-twentieth century led to the emergence of a new dispute just west of the strait in the Pacific Ocean. The dispute centres on Canada's straight baselines, which it adopted along the indented southwest coast of Vancouver Island in 1969. The USA immediately objected on the grounds that the baselines had been constructed 'contrary to established principles of international Law of the Sea' (Byers and Østhagen 2017, 23).

There is no evidence of pressure from the fishing industry to resolve the dispute. As in the case of the Dixon Entrance, the cooperative management of the fishery under the Pacific Salmon Commission, combined with flag-state enforcement, has created a situation that is workable for both sides (McRae 1989, 154–55). For this reason, public opinion does not play a role: indeed, very few Canadian or US citizens are even aware of the existence of the dispute. There is some degree of regional interest, however, with the province of British Columbia expressing the view in the 1970s that the boundary should follow the underwater Juan de Fuca Canyon, not an equidistance line (British Columbia 1977).

As in the other Canada–USA boundary disputes, both countries seem concerned that compromising on a principle of delimitation in one instance could weaken their position in another. Moreover, the same concern may exist over the law governing straight baselines. The Canada–USA dispute seaward of Juan de Fuca Strait could in fact be linked to a dispute over straight baselines in the Arctic. When Canada adopted straight baselines around its high Arctic archipelagos in 1985, it brought immediate protests from the USA and the European Community (EC – which became the EU) (Byers 2013, 133–34, 137–38). Therefore, both Canada and the USA might be concerned that any compromise on straight baselines along Vancouver Island could weaken their positions in the Arctic, where the dispute over straight baselines is linked to the far more significant dispute over the status of the Northwest Passage.

5.2   DENMARK (GREENLAND)

In 1970, Canada extended its territorial sea from 3 to 12 n.m. (Government of Canada 1970). However, it overlooked the fact that at several points, the new limit extended more than halfway across Nares Strait, the narrow channel between Ellesmere Island and Greenland (Gray 1997, 68). Once this consequence was realised, boundary negotiations with Denmark commenced. The boundary under negotiation was potentially quite extensive because Greenland lies within 400 n.m. of the long eastern coastlines of both Ellesmere Island and Baffin Island, each of which is larger than the UK.

In 1973, Canada and Denmark agreed to divide the ocean floor using an 'equidistance line' (Canada–Denmark 1973). Since then, the two countries have also used the resulting 1,450 n.m. boundary to define their fishing zones: the continental shelf delimitation has become an all-purpose maritime boundary (Gray 1997, 68). Article V in the 'Agreement on the Continental Shelf between Greenland and Canada' addresses the possibility of hydrocarbon reserves straddling the new boundary and proposes that the parties find ways to mutually exploit the resources (Canada–Denmark 1973).

The treaty does have one unusual element: how it deals with a disputed island located on the equidistance line. Hans Island, with an area of only 1.3 km², is not mentioned in the treaty (Byers 2013, 10–16). The maritime boundary stops just short of the south shore of the island and begins again just off the north shore. As a result, the dispute over Hans Island has been rendered almost irrelevant because it now concerns only a tiny piece of land, with the surrounding seabed and water column having been allocated by treaty (and practice consistent with that treaty). As Oxman (1995, 250) explains, 'Canada and Denmark are said to have been motivated by the desire to avoid future disputes in a largely unsettled area where Greenland faces the Canadian Arctic'. Similarly, Alexander (Alexander 1993, 372) writes, 'The agreement shows a strong effort by both parties to avoid conflict in marine resource exploitation.'

The Lincoln Sea is the portion of the Arctic Ocean located directly to the north of Greenland and Ellesmere Island. The thickest sea ice in the Arctic is found there, pushed into the space between the two landmasses and held there for years by prevailing winds and ocean currents. In 1973, the negotiators who delimited the maritime boundary between Canada and Greenland stopped at 82°13' north, where the Nares Strait opens into the Lincoln Sea. Then, in 1977, Canada claimed a 200 n.m. fisheries zone along its Arctic Ocean coastline. The zone was bounded in the east by an equidistance line that used the low-water mark of the coasts of Ellesmere Island and Greenland and several fringing islands as its base points (Gray 1997, 68).

Three years later, Denmark adopted its own equidistance line, but only after drawing straight baselines – two of which used Beaumont Island as a base point. In 1982, Canadian and Danish diplomats met to discuss the Lincoln Sea boundary dispute, 'with neither side moving from their respective positions' (Calderbank et al. 2006, 163). In 2004, the scope of the dispute was reduced when Denmark modified its straight baselines (UN Law of the Sea 2005). These developments may have contributed to the announcement by the Canadian and Danish foreign ministers in 2012 that negotiators 'have reached a tentative agreement on where to establish the maritime boundary in the Lincoln Sea' (Canadian Department of Foreign Affairs 2012; Mackrael 2012). In 2018, the two countries further established a 'Joint Task Force on Boundary Issues' to settle the outstanding issues regarding the maritime boundary (Global Affairs Canada 2018).

5.3   FRANCE

St. Pierre and Miquelon is an archipelago of eight islands with a total landmass of 242 km². Located just 13 n.m. from the coast of Newfoundland, the islands were claimed by Jacques Cartier on behalf of France in 1536. The islands changed hands several times during wars between France and Britain but have remained uncontested French territory since 1815. The islands support a population of around 6,000 people, with an economy based on fishing and tourism.

The dispute over the maritime zones around St. Pierre and Miquelon began in 1966, when the Canadian and French governments exchanged diplomatic notes setting out their positions with respect to the delimitation of the continental shelf (Canada–France 1992). The exchange of views was prompted by both countries granting oil and gas exploration licences in the area (Canada–France 1992). In 1972, the two countries concluded a maritime boundary treaty resolving overlaps within 12 n.m. of the coasts of Newfoundland, on the one hand, and St. Pierre and Miquelon, on the other hand (Canada–France 1972). Canada and France then spent years negotiating an extension of the boundary out to 200 n.m. (the EEZ) before agreeing in 1989 to send the matter to an ad hoc arbitral tribunal.

In 1992, the tribunal issued a highly unusual decision (Canada–France 1992).[2] It awarded France a 24 n.m. wide band around the seaward side of the islands, along with a 10.5 n.m. wide corridor extending 188 n.m. southwards from the islands. If the corridor had been intended to allow France access to its territorial sea and EEZ without having to pass through Canada's EEZ, it failed to accomplish this because Canada's zone extends farther offshore, around the stem of the mushroom-shaped French zone (Byers and Østhagen 2017, 32). Fisheries provided the principal motivation for the negotiations and the eventual recourse to third-party dispute settlement (Saunders and VanderZwaag 2010, 209). Hydrocarbons had already been discovered on either side of the disputed zone, and as noted, the two countries had independently issued overlapping exploration licences in the zone itself.

5.4   IN THE SHADOW OF THE HEGEMON

Most of Canada's maritime boundary disputes remain unresolved or are only partially resolved. In the cases of the Lincoln Sea, Machias Seal Island and seaward of Juan de Fuca Strait, the resources located within the disputed zones are speculative, commercially unviable or relatively small. There is considerable hydrocarbon potential in the Beaufort Sea, but this has not been realised because of the high operating costs and the easier availability of comparable resources elsewhere. In the Dixon Entrance, Canada and the USA have worked out an arrangement allowing fishers from each side to access the disputed zone, here subject to flag-state enforcement.

Significantly, although negotiations on the Beaufort Sea boundary were initiated after oil prices rose, they were suspended when prices fell. In the Gulf of Maine and around St. Pierre and Miquelon, relatively high levels of economic activity and the potential for a 'cod war' scenario involving repeated and reciprocal arrests of fishing boats eventually pushed the disputing parties into adjudication and arbitration. In the Beaufort Sea, uncertainty about the existence and location of hydrocarbons played a role. After initiating boundary negotiations with the USA in 2010, uncertainty concerning the existence and location of hydrocarbons seems to have contributed to the suspension of the discussion. Efforts were made to resolve the uncertainty through seismic mapping of the disputed zone, but the resulting delay coincided with a change of Canadian foreign ministers and a sharp drop in world oil prices.

One central notion here, as in the case of Canada, is the difficulties of reaching an international agreement acceptable to a domestic audience when the issues have become politicised: that is, when issues reach the policy agenda of certain actors potentially holding contrarian positions. When economic interests required a settlement, as occurred in the Gulf of Maine and around St. Pierre and Miquelon, Canada worked towards a boundary resolution – in both cases by outsourcing the actual drawing of the line to objective and disinterested third parties.

Canada has also been cautious about compromising on its legal position in the Beaufort Sea because of concerns that this might detrimentally affect its position on other boundary disputes. This is why it sought a 'package deal' in 1977 (Kirkey 1995, 58–59). In 1977–78, Canada and the USA found themselves in a zero-sum negotiating situation in the Beaufort Sea: either Canada would have to surrender on the 141st meridian, or the USA would have to surrender on the equidistance principle. Concerns about precedents made these options even less palatable. Canada was seeking a way out of the zero-sum scenario when it proposed a package deal – one that, for instance, could have allowed a US 'win' in the Beaufort Sea in return for a Canadian 'win' in the Gulf of Maine. If Canada could resolve all four disputes with the USA simultaneously, its concerns about a precedent would disappear. That was not the case with the USA, however, because of US concerns about setting a precedent with Canada would extend to disputes with other countries.

Further, there can be little doubt that Canada's key concern in connection with its maritime boundaries is its relations with its powerful southern neighbour, here moving beyond only concern for legal precedent (McDorman 2009). This is again where the notion of regional and local patterns of interaction come into play. In the other boundary instances – with France and Denmark/ Greenland – solutions, albeit tricky and time-consuming, were eventually found (Byers and Østhagen 2018). It is with the USA that all of Canada's outstanding boundary disputes remain: this speaks to the complexities of this relationship. This relationship is intricate on its own – and the USA, with its resources and leverage, is no easy negotiating partner (Ásgeirsdóttir 2016), even for close allies.[3]

Several other features of Canada and its maritime boundaries, settled as well as unsettled, should be taken into account. Canada's federated system makes it complicated for any Ottawa-based government to conclude negotiations without consent from the regional and local levels, especially when provinces are involved. Some of Canada's maritime boundary disputes are also based on historic treaties or claims, in contrast to a 'simple' boundary dispute that arises when two states expand their zones based on previously settled borders/ boundaries.

Furthermore, some factors are not evident in Canada's maritime boundary disputes. Oil and gas activity, real or potential, seems modest. Security concerns are almost completely absent, even in relation to small-scale issues such as IUU fisheries and clashes between fishers from opposing countries. A pure geographic fact – namely remoteness and limited access to some of the maritime domains in question, especially concerning the Arctic – has also figured in the limited attention devoted to maritime boundary settlement until only recently. Finally, despite Arctic involvement under the Harper government starting in 2006, the maritime domain is not central to Canada's policy thinking, nor to its economic outlook. The absence of these factors also helps to explain why so many of Canada's maritime boundaries still remain disputed.

NOTES

1. Kirkey (1995, 59–60) argues, 'U.S. officials were concerned that by deviating from this position, which seeks to delimit wet boundaries according to the principle of equidistance – except in cases where specifically defined circumstances exist – American ability to successfully prevail either in the course of international negotiations over future maritime boundary cases, or regarding those cases brought before the ICJ, would be greatly reduced'.
2. See McDorman 2009; Gray 1997; Cook 2005; and Byers and Østhagen 2017.
3. See McDorman 2009; Gray 1997; Cook 2005; and Byers and Østhagen 2017.

6. Colombia – through the Caribbean labyrinth[1]

As the international legal regime for the oceans developed from the 1950s and onwards, Latin American countries were vigorous in establishing extended maritime zones. An approach emerged where the settlement of maritime rights preceded economic activity or conservation was prevalent among these countries: they were among the first in the world to establish the 'patrimonial sea', which later became the EEZ (Nweihed 1980; 1993f, 275). Colombia was also party to the American Treaty on Pacific Settlement, signed by the independent republics of America in Bogota, Colombia, on 30 April 1948. Known as the Bogota Pact, 21 countries signed it to establish channels of communication for territorial disputes, among other things. Colombia, however, withdrew in 2012, following the ICJ decision in the maritime dispute with Nicaragua (BBC News 2012).

Colombia approached its maritime neighbours in the 1960s and 1970s to settle boundaries before any serious disputes could emerge. In the late 1960s, negotiations started with Venezuela over the presumably oil-rich north-western corner of the Gulf of Venezuela (Nweihed 1980). Negotiations were also initiated with Nicaragua, though the related maritime boundary concerns the larger issue of the 1928 Esguerra–Bárcenas Treaty and the subsequent dispute over the San Andrés Archipelago and its surrounding waters, which was eventually brought before the ICJ. That situation is still unresolved. However, in the 1970s, Colombia managed to settle several boundaries: with Ecuador (1975), Panama (1976), Costa Rica (in the Caribbean, 1977), the Dominican Republic (1978) and Haiti (1978). It was not until 1978 that Colombia claimed its full EEZ through a legal act – which was somewhat late in the South American context (Government of Colombia 1978).[2]

Then, after a decade of negotiations, the maritime boundaries with Costa Rica (in the Pacific, 1984), Honduras (1986) and Jamaica (1993) were concluded. Here, it should be noted that some of these boundary agreements were not ratified until later because of domestic political considerations in one or both of the respective countries; the 1977 agreement with Costa Rica has still not been ratified. In all these negotiations and – when applicable – agreements, an 'all-purpose' boundary was employed by delineating the continental shelf,

Source: https://en.wikipedia.org/wiki/File:COL_orthographic_(San_Andr%C3%A9s_and
_Providencia_special).svg.

Figure 6.1   *Map of Colombia*

economic zones, fishery zones and possible other zones (Nweihed 1993f, 276).
(See Figure 6.1.)

6.1   ECUADOR

As LOS proceedings seemed to be solidifying around a 200 n.m. economic
zone, Colombia began to realise that negotiating maritime boundaries would
become increasingly important (Londoño 2015, 247). It was recognised that
a dispute might ensue in tandem with preliminary talks with both Nicaragua
and Venezuela in the late 1960s and early 1970s. However, some of
Colombia's regional neighbours with a Pacific coastline were among the first
countries in the world to expand their maritime zones up to 200 n.m.: Chile,
Peru and Ecuador all declared extended zones to protect national fisheries.
In 1952, delegations from the three countries met in Santiago and signed the
Declaración sobre Zona Marítima ('the Santiago Declaration'), in which they
proclaimed 200 n.m. zones to 'assure their people the necessary conditions of
subsistence and to procure the means for their economic development' (Chile,
Ecuador, and Peru 1952).

The border region between Colombia and Ecuador had experienced some instability in terms of unlicensed trade and unregistered movement of people because of the limited state presence in this inhospitable terrain. However, '[t]he countries had similar approaches to maritime delimitation, and never had any serious issue arise between them. Both countries therefore agreed to take a first step in avoiding future problems by agreeing on a maritime boundary sooner rather than later' (Londoño 2018). Subsequently, Ecuador became the first country with which Colombia settled a maritime boundary.

The boundary agreement between Ecuador and Columbia was signed on 23 August 1975 and entered into force on 22 December of that year (Colombia–Ecuador 1975). Afterwards, showcasing the limited public conceptualisation of maritime boundaries at the time, the Colombian government was criticised for the futile effort of drawing an 'imaginary line in the sea' (Londoño 2015, 248). As put by Jiménez de Aréchaga (1993, 810–11), '[T]he agreement refers to preservation, conservation and utilisation of resources in the marine and submarine areas, although this does not seem to have had any effect beyond the statement, i.e. in special provisions or the shape of the boundary line'.

6.2   PANAMA

One year after the agreement with Ecuador, Colombia concluded the first agreement in South America involving a 'double frontage' – in two maritime domains at once. This concerned the larger sovereignty dispute between Colombia and Nicaragua, in which the boundary defined in the Caribbean would be influenced by the outcome of that dispute. Thus, it fell 'within the strategy set by Colombia in the 1970s aimed at completing its maritime map' (Nweihed 1993e, 520).

The maritime boundary dispute with Colombia concerned the maritime areas related to the Panama Canal, in turn making it strategically significant for not only Colombia and Panama, but also other states concerned with the canal. In the 1970s, Panama focused on nationalising the Panama Canal, which had been under US jurisdiction since 1904. In 1977, a treaty settled the status of the canal, which was to be completely returned to Panama by 1999 (United States–Panama 1977). In this process, Colombia had signalled to both Panama and the USA that Colombian rights regarding the canal would have to be respected and upheld (Londoño 2015, 252).

The Treaty between Colombia and Panama was signed on 20 November 1976 (Colombia–Panama 1976) and entered into force on 30 November 1977. The text specifically refers to the good neighbourly relations between the two countries and the importance of international cooperation. Additionally, mention is made of the 'resources which exist in the waters', though no joint development zones were constituted through the treaty. The parties emphasised LOS principles such as freedom of navigation, innocent passage and free traffic, in line with concerns regarding the Panama Canal (Nweihed 1993e, 523; Colombia–Panama 1976). The agreement encountered some domestic opposition in Panama because it was seen as favouring Colombia. In the end, the treaty was ratified by both countries, due in part to the authoritarian regime in Panama at the time putting pressure on the Panamanian Congress (Nweihed 1993e; Londoño 2015, 256).

6.3 COSTA RICA

One year after the agreement with Panama, Colombia and Costa Rica negotiated a maritime boundary in the Caribbean. This is one of the two domains (the other being in the Pacific) where the Costa Rican and Colombian EEZs meet. Although the two countries have no land border, Colombian sovereignty over islands in both the Pacific (Malpelo) and the Caribbean (the Archipelago of San Andrés) led to the need for maritime delimitation. Efforts with Costa Rica also followed Colombia's systematic attempts to settle its maritime boundaries in the late 1970s as the third LOS conference proceeded (Nweihed 1993a, 463).

The treaty concerning both the boundary and maritime cooperation was signed on 17 March 1977 (Colombia–Costa Rica 1977). As with Colombia's previously settled maritime boundaries (Ecuador and Panama), the delimitation concerned a 'total boundary' including both an EEZ boundary and continental shelf boundary. The economic incentives for the treaty were seen as being limited at the time, despite there being some fisheries in the relevant maritime domain (Nweihed 1993a, 469). The line itself was based on equidistance, while deviating at times to compensate both countries on various positions.

The treaty links directly with Colombia's dispute with Nicaragua, explicitly recognising the full maritime zone of the San Andrés Archipelago up to the 82nd meridian. Costa Rica's acquiescence to this boundary agreement was seen as taking sides in the dispute between Colombia and Nicaragua (Londoño 2015, 257). The Costa Rican government delayed, and then in 1983, the ratification process was withdrawn. In part, this was due to strong domestic opposition and lobby efforts against the treaty, where interest groups questioned its usefulness and its potentially ruinous effects on relations with neighbouring Nicaragua (Nweihed 1993a, 464).

In 1984, the two countries negotiated their second maritime boundary in the Pacific. The two boundaries were initially combined for the Costa Rican ratification process, here without success. The agreement was signed on 6 April 1984. The main impetus for this second agreement was protection of national fisheries in both countries, Costa Rica in particular (Aréchaga 1993, 810). The agreement delineates the maritime boundary between Isla del Coco

(Costa Rica) and the Island of Malpelo (Colombia). As with the treaty in 1977, the 1984 agreement encountered domestic opposition in Costa Rica. It took considerable political handiwork and several governments to get it ratified in 2001 – only after it had been decoupled from the 1977 agreement. On 20 February 2001, the 1984 Pacific boundary was ratified in both countries, but (as mentioned) the 1977 Caribbean boundary has remained unratified due to the disagreement with Nicaragua (Charney and Smith 2002). However, the two countries operate with the 1977 boundary as the de facto boundary (Londoño 2015, 261).

6.4   DOMINICAN REPUBLIC

In 1978, Colombia and the Dominican Republic agreed on a single maritime boundary in the Caribbean. The agreement concerning both maritime delimitation and cooperation was signed on 13 January 1978 (Colombia–Dominican Republic 1978). As with Colombia's three previous agreements, the emphasis was on bilateral cooperation under the framework of LOS to protect regional and national fisheries (and other related resources).

Nweihed (1993b, 480; 1983) holds that this agreement was more important to the Dominican Republic because of its dependence on regional fisheries in the area and was less important – from an economic point of view – for Colombia. However, it holds another form of significance for Colombia: the maritime boundary relates to the second of Colombia's complicated maritime relationships, here with Venezuela in the Gulf of Venezuela (Londoño 2015, 265). Whereas Points 1 and 2 in the boundary itself are established primarily by means of equidistance between Colombia and the Dominican Republic, Point 3 (the final point of the boundary) depends on the location of a hypothetical maritime boundary between Colombia and Venezuela, which has yet to be determined.

The agreement between Colombia and the Dominican Republic led to protests by the Venezuelan Deputy Foreign Minister (Nweihed 1980, 11–12). Venezuela also put considerable diplomatic pressure to bear on the Dominican Republic through official letters and private negotiations (Londoño 2015, 267). Complicating matters further, in 1979, the Dominican Republic had agreed its maritime boundary with Venezuela, partly ignoring the agreement with Colombia. The boundary between Colombia and the Dominican Republic remains in place, though its final part is still contingent on an agreement with Venezuela.

6.5   HAITI

Following the four previous maritime boundary agreements, Colombia continued its quest to settle all related EEZ boundaries in the Caribbean and Pacific. As with the Dominican Republic, Colombia's EEZ (deriving from its Caribbean coast) meets the EEZ of the Caribbean Island of Haiti. By concluding this delimitation, Colombia also sought to solidify its boundaries and develop a case vis-à-vis Venezuela (Nweihed 1993c, 491). Both Colombia and Haiti considered themselves allies of the USA, which had a considerable naval presence in the region at the time (Nweihed 1993c, 492).

The agreement was signed on 17 February 1978 and came into force on 16 February 1979 (Colombia–Haiti 1978). The provisions of the agreement focus on avoiding marine pollution and taking measures to protect migratory (marine) species, such as fish stocks. The two countries declare, 'Close collaboration is essential for the preservation, conservation and use of existing resources' (Colombia–Haiti 1978).

6.6   HONDURAS

Colombia had seized some of the 'low-hanging' opportunities in the Caribbean and Pacific, but a few difficult maritime boundaries remained. The boundary with Honduras in particular had the potential to cause public concern and international tension because of its link to the dispute with Nicaragua. After Nicaragua had granted (unused) exploratory hydrocarbon licences on the continental shelf east of the 82nd meridian in 1967, Colombia had been exchanging notes with Honduras since 1970 concerning the various islands and cays around San Andrés (Nweihed 1993d, 505). Officially, however, negotiations started in 1982, when in a surprise move, Honduras included the (Colombian) bank of Serranilla as a part of its territory in a constitutional reform (Londoño 2015, 271). Moreover, in 1982, the new President of Colombia, Belisario Betancur, initiated a turn in Colombian foreign policy, which coincided with a new US approach towards Nicaragua and Latin America to combat rising drug trafficking to the USA.

In addition to Colombia's desire to bolster its case against Nicaragua, it was thought that the maritime domain in question held considerable oil and gas resources, a point that gave economic impetus to the negotiations (Nweihed 1993f, 279). Moreover, the area in question was important for Honduran fishers. By 1978, Honduras had some 260 fishing vessels and was slowly expanding its fleet to catch shrimp, pelagic stocks and lobster (Nweihed 1993d, 509). These various factors opened a policy window for the two states, which spent several years, starting in 1982, negotiating an agreement.

The treaty was signed on 2 August 1986, though it did not enter into force until 20 December 1999. The treaty explicitly notes the 'friendship bonds that rule their relationship' (Colombia–Honduras 1986). Concerning oil and gas resources, the treaty sets provisions for the unitisation of hydrocarbon and natural gas development to ensure equal exploitation of potentially straddling resources (Colombia–Honduras 1986). Importantly, this agreement recognises Colombian sovereignty, not only over the island of San Andrés and the Archipelago, but also the various banks and cays located north of the Archipelago. It provides explicit recognition of the Colombian interpretation of the treaty from 1928 between Nicaragua and Colombia, in which the boundary is based on the 82nd meridian (Londoño 2015, 277; Nweihed 1993d, 509).

The agreement caused reactions in several countries, first and foremost Nicaragua, which protested against the bilateral decision and what it saw as infringement of Nicaraguan sovereignty (Londoño 2015, 248). There was domestic opposition in Honduras as well. The main concern was that giving away the Serranilla Bank was contrary to Honduras' 1982 constitution regarding territorial cohesion (Sandner and Ratter 1991, 299). Further, Colombia heightened its naval presence in the area and actively prevented Honduran fishermen from accessing it. There were additional fears that Honduras had conceded too much to Colombia. Although equidistance had guided most of the boundary-making, there was a deviation for the Serranilla Bank, where 75% of its economic potential was ceded to Colombia, as well as the entire Bajo Nuevo Bank (Nweihed 1993d, 513).

In Colombia, there was further opposition to the agreement. Some argued it was contradictory to Colombia's commitment to the peaceful settlement of disputes with other Latin American countries, as per the Bogota Pact. There was also controversy concerning the installation of a US military radar base on the island of San Andrés, which, in the end, was rejected by Colombia in 1986. The radar base had been intended to help fight drug trafficking, but also to monitor activities in Nicaragua after the left-wing revolution there in 1979 (Sandner and Ratter 1991, 300; Nweihed 1993d, 507). The Colombian Senate, however, approved the maritime boundary agreement only two months after its signing on 23 October 1986. Although the agreement was signed – symbolically, on the island of San Andrés – in 1986 and ratified in Colombia two months later, it did not come into force until 1999, when it was ratified by Honduras.

6.7   JAMAICA

Jamaica had initially been one of the states opposed to the growing 'territorialisation' of the maritime domain through the codification of sovereign rights extending all the way out to 200 n.m. in the 1960s and 1970s. In the Santo

Domingo Conference on the Law of the Sea in 1972, Jamaica voted for 'matrimonial sea', that is, regional seas instead of individual seas appropriated to one state (Nweihed 1998, 2183). This did not, however, gain support in the face of 'advancing state jurisdiction' (Rattray, Kirton, and Robinson 1974).

The 1993 agreement with Colombia was the first boundary agreement to be concluded by Jamaica. The treaty was signed on 12 November 1993 (Colombia–Jamaica 1993) and entered into force only four months later on 14 March 1994. The agreed-upon maritime boundary entered the picture with Colombia's previously agreed boundaries with Haiti, Honduras and the Dominican Republic. It also connected with Colombia's dispute with Nicaragua, affirming Colombian sovereignty over the Serranilla Bank and Bajo Nuevo.

One dimension of this maritime boundary agreement that has received considerable attention is the creation of a 'Joint Regime Area' (JRA). The JRA is a triangular zone between the two countries stretching from Serranilla Bank and Bajo Nuevo Bank in the north towards a specific point further south. Although the treaty affirms Colombian sovereignty over the structures above water, the JRA is 'internally undelimited' (Colombia–Jamaica 1993, 2183). The JRA specifies several joint activities in which the states may engage, ranging from establishing artificial islands to pursuing economic ventures.[3]

Regarding the incentives for agreeing beyond the factors already mentioned, access to important fisheries for Jamaica seems to have been a priority (Londoño 2015, 291). Colombia negotiated and signed a fisheries agreement with Jamaica in 1982, which went in force starting in 1984, which occurred at the same time as negotiating with Honduras. Although an agreement was reached with Honduras in 1986, the lack of ratification might have led Colombia to pursue a broader agreement with Jamaica. Pressure from Jamaican fishermen to gain access to these waters also contributed to the eventual outcome: the JRA.

6.8   NICARAGUA

In 1928, Colombia and Nicaragua concluded the Esguerra–Bárcenas Treaty, and later, in 1930, by exchange of notes (known as the Managua Treaty), agreed on a border along the 82° W meridian. As noted, this inferred sovereignty over the Intendencia de San Andrés y Providencia to Colombia (Nieto-Navia 2015; Sandner and Ratter 1991). In 1967, Nicaragua granted oil exploration licences around Quitasueño Bank – east of 82° W, leading to diplomatic protests from Colombia (Nweihed 1993e, 524). Colombia followed up by initiating negotiations over the issue in 1969 (Londoño 2018).

By 1980, Nicaragua had denounced the whole agreement, arguing that it should enjoy full sovereignty over both the San Andrés Archipelago and related islands and reefs, as well as the maritime zones. Colombia responded with a white paper arguing its sovereignty, consequently stepping up its military presence in the area (Sandner and Ratter 1991, 299). The sovereignty of the islands and the sovereign rights in the waters around the San Andrés Archipelago have since remained disputed. Colombia has maintained a naval presence, as well as presence on the various islands and cays. In 1991, the intendencia was elevated in Colombia's new constitution to the status of a Department (Palacios 2006).

On 6 December 2001, Nicaragua initiated proceedings against Colombia at the ICJ, requesting that the court declare Nicaraguan sovereignty of the islands of Providencia, San Andrés and Santa Catalina, in addition to nearby islands and keys like Serranilla (ICJ 2018). On 19 November 2012, the court held that by virtue of the 1928 treaty between Colombia and Nicaragua, Colombia had sovereignty over the islands and reefs belonging to the San Andrés Archipelago (ICJ 2012). The question then became how to define where this jurisdiction ends, geographically.

The court concluded that Colombia had sovereignty over the islands/cays of Albuquerque, Bajo Nuevo, East–Southeast Cays, Quitasueño, Roncador, Serrana and Serranilla (ICJ 2012). Concerning the maritime boundary, however, the court rejected the original 82° W meridian and awarded Nicaragua the full 200 n.m. EEZ, with only limited maritime zones attributed to the Archipelago of San Andrés and the other islands, cays and reefs. The court could not, however, rule on Nicaragua's request for an extended continental shelf out to 350 n.m. because sufficient technical evidence had not been provided.

In total, the verdict transferred some 75,000 km^2 of maritime area to Nicaragua, an area rich in fisheries and thought to hold hydrocarbon deposits (Economist 2012). The ruling turned the uninhabited Colombian islands of Quitasueño and Serrana into isolated enclaves, leaving the rest of the San Andrés Archipelago extending into Nicaragua's newly expanded maritime area. Colombia responded by rejecting the decision. President Santos refused to withdraw the Colombian Navy and argued that the ruling was full of 'omissions, mistakes, excesses [and] inconsistencies that we cannot accept' (Economist 2012). Nine days later, Colombia withdrew from the Pact of Bogota, through which it had pledged to resolve disputes through the ICJ. Santos declared, 'Never again should we have to face what happened to us on November 19th [the date of the ICJ verdict]' (Economist 2012).

Lawyers and scholars engaged with the LOS, however, were of another opinion (Galvis and Arévalo 2018). Colombian legal scholars who were close to the government and the preparatory work for the case itself held that the view that Colombia had 'lost' was based on a misconception. The 82nd meridian had never been the actual maritime boundary between the two states, as per international law. The problem was that it had been marketed domestically in Colombia as a given (Londoño 2018). The dispute is still unresolved.

6.9   VENEZUELA

When Colombia and Venezuela gained independence from Spain in the period of 1810–1819, they were initially joined under the structure of Gran Colombia (until 1831). When this federation dissolved and Venezuela became an independent country, the precise border on land in the Gulf of Venezuela remained undefined. Despite attempts to settle the land border in the nineteenth century, the final border was not agreed upon until 1941, with the Lopez de Mesa–Gil Borges Treaty that set the demarcation of the Colombia–Venezuelan border (Colombia–Venezuela 1941).

However, Venezuela claims the Gulf of Venezuela as historic waters, considering the gulf as closed by a line drawn in 1939 across its mouth from the Castilletes point where the land boundary ends. Also, the status of the Los Monjes islets is disputed, with Colombia holding that they belong to Colombian territorial seas, whereas Venezuela considers them to be an extension of its continental shelf (Nweihed 1980, 9). Initially ceded to Colombia by Venezuela in a diplomatic note in 1952, the decision was never ratified by the Colombian Congress (Londoño 2015, 220–21).

In 1975, the presidents of Colombia and Venezuela submitted to their respective congresses a proposal for a resolution of the issue. After several rounds of negotiations, the agreement with Venezuela failed to achieve support in the Venezuelan Congress and was rejected (Nweihed 1993f, 282; Londoño 2015, 235). The dispute peaked in 1987, when Colombia positioned two naval vessels in the disputed waters. Venezuela reacted by launching naval vessels and fighter aircraft towards Colombia and massing ground troops along the border (Tovar 2015). The risk of conflict was resolved by diplomacy and the Colombian vessels leaving the waters. In the 1990s, relations between the two countries improved, and bilateral cooperation was expanded to trade, antiterrorism and border affairs (LaRosa and Mejía 2012). Relations took a negative turn, however, when Hugo Chavez took over the Venezuelan presidency in 1999. Negotiations on the maritime boundary were then blocked in 2009, when the Chavez government dissolved the boundary negotiation commission (Tovar 2015).

Other incursions by military units from both countries have led to diplomatic protests and sometimes even military posturing (Tovar 2015; Caro 2017). However, the maritime boundary between the two countries does not top their bilateral agenda (Casey and González 2019).

6.10   THROUGH THE CARIBBEAN LABYRINTH

Colombia's first boundary with Ecuador led to the government being criticised for drawing an 'imaginary line in the sea'. At the other end of the spectrum, when Colombia agreed its final (to date) maritime boundary with Jamaica in 1993, some domestic actors saw the agreement as 'giving away' maritime space that was supposedly Colombian. As Londoño (2015, 296) notes, it was once a widely held belief that everything in the Caribbean northwards of Colombia's mainland was Colombian. In 20 years, the public discourse concerning maritime space had changed.

For the other parties in Colombia's maritime boundary agreements, economic opportunities appear to have had motivating effects, to varying degrees. With Ecuador, Costa Rica, the Dominican Republic, Haiti, Jamaica and Honduras, access to fisheries that were already important to local fishermen (and, to some extent, the governments) have been a motivating factor, though Colombia had fewer stakes in fisheries than did its negotiating partners regarding the various Caribbean islands. Oil and gas prospects seem to have been involved in the case of Honduras, albeit not as the primary reason for settlement, and as a factor in driving negotiations with Venezuela.

One approach favoured by Colombia to alleviate concerns about losing valuable resource-access has been the use of joint zones/regimes, their characteristics varying with each boundary agreement. Colombia was willing to employ such mechanisms quite early in the international context as a way of achieving boundary delimitation, despite uncertainty over where potential resources might be located.

Engagement with domestic interests was evident in the proceedings after the agreement with Panama, in which domestic opposition in Panama criticised as being too favourable to Colombia; with Costa Rica, where fears of damaging relations with Nicaragua were voiced by the opposition in Columbia; with the Dominican Republic, where similar fears of the impact on relations with Venezuela were brought forward; with Honduras, where fear of losing rights and damaging relations with Nicaragua came to the fore; and with Jamaica, where fishing interests in both countries voiced their concerns.

There remain only two disputes, both of them complex and difficult. In the early 1970s, Colombia instigated negotiations with both Nicaragua and Venezuela to scope out the possibilities for a maritime boundary agreement, although these two disputes have proven the hardest nuts to crack. This further helps explain Colombia's strategic approach to maritime boundary dispute settlement. Colombia intentionally sought to build its case through legal precedents against Nicaragua's claims, but the ICJ verdict in 2012 did not favour the Colombian position (Tanaka 2013). However, Colombia managed to affirm sovereignty over not only the San Andrés Archipelago, but also the surrounding islands and cays, as well as receiving affirmation of limited maritime zones.

Thus, the case of Colombia shows how one state can take on a highly active approach to settling its maritime boundary disputes with an eye towards a larger geopolitical and legal goal. Whether Colombia can be deemed successful in the dispute with Nicaragua is debatable. Colombia did, however, manage to settle all its other maritime boundaries – except for Venezuela – in a complex web across the Caribbean and Pacific.

NOTES

1. 'Labyrinth' is borrowed from Londoño 2015.
2. Efforts by Colombia and the UK concerning a possible boundary between the Serranilla Bank and related waters and the British Grand Cayman Islands will not be discussed here because of the very limited scope of the boundary and negotiations (Londoño 2015, 287–88).
3. Colombia–Jamaica Treaty 1993: 'In the Joint Regime Area, the Parties may carry out the following activities: (a) Exploration and exploitation of the natural resources, whether living or non-living, of the waters superjacent to the seabed and the seabed and its subsoil, and other activities for the economic exploitation and exploration of the Joint Regime Area; (b) The establishment and use of artificial islands, installations and structures; (c) Marine scientific research; (d) The protection and preservation of the marine environment; (e) The conservation of living resources; (f) Such measures as are authorized by this Treaty, or as the Parties may otherwise agree for ensuring compliance with and enforcement of the regime established by this Treaty'.

7. Norway – looking to Russia and the Arctic

The northern European countries around the North Sea – Norway, the UK, the Netherlands and Denmark – were among the first to push for an extension of maritime zones in the postwar period, during a time when this was becoming accepted around the globe.[1] In addition to the North Sea, Norway has access to the Norwegian Sea and the Barents Sea, as well as the Arctic Ocean through the Svalbard Archipelago in the north (Figure 7.1).

As early as 1812 (when under Danish rule) Norway had established 4 n.m. territorial waters from the outermost limit of its many islands and reefs (Riste 2005, 63). In the 1860s, Norway drew a straight baseline across the Vestfjord (between the city of Bodø and the Lofoten Archipelago) to protect local fisher-

Source: https://en.wikipedia.org/wiki/Norway#/media/File:Europe-Norway_(orthographic _projection).svg.

Figure 7.1   Map of Norway

ies against French fishers, which caused an outcry in France (Riste 2005, 63). By 1935, Norway had established straight baselines more generally north of the Arctic Circle, which aroused the irritation of British fishers who objected to being excluded from Norwegian waters (Harsson and Preiss 2012). In the end, the UK took the case to the ICJ, which in 1951, with the Anglo-Norwegian fisheries case, upheld the Norwegian approach regarding straight baselines (Green 1952).

In the postwar period, Norway placed emphasis on its maritime domain as a source of fisheries and shipping – historically the backbone of its economy. Norway was also among the countries taking the lead in efforts at the UN to develop a comprehensive approach to the oceans, which led to the Geneva Convention in 1958 and later the LOS (Retzer 2017). The 'Evensen group' – named after Norwegian Minister of Maritime Affairs Jens Evensen – became central in finding compromises during LOS negotiations in the 1970s (Retzer 2017; Harsson and Preiss 2012). Norway was also active in pushing for an extension of the continental shelf beyond 200 n.m. (Ø. Jensen 2014a, 41), which was later enshrined in LOS Article 76 (1).

Negotiations between Norway and its maritime neighbours on how to delineate maritime space began in the 1960s when the oil and gas potential of the North Sea became apparent. This move provided the impetus to delimit Norway's seabed boundaries with the UK and Denmark in areas of the North Sea that had previously been considered the high seas. A dispute also arose in the 1960s when both Norway and the Soviet Union drew on the 1958 Geneva Convention on the Continental Shelf in claiming offshore rights (Henriksen and Ulfstein 2011). The dispute in the Barents Sea with the USSR expanded in geographical scope when Norway established its 200 n.m. EEZ in 1976 and the USSR established its zone in 1977.

Having established an EEZ along its entire coast in 1976, Norway also decided to establish a fisheries protection zone (FPZ) around the Svalbard Archipelago in 1977, arguing that the 200 n.m. maritime zone around Svalbard is not subject to the Spitsbergen Treaty (Ø. Jensen 2014a, 102). To avoid an outright challenge to the Norwegian claim and protect and manage what is the central spawning area for the northeast Atlantic Cod stock, the Norwegian government opted to establish the FPZ instead of an EEZ (Jørgensen and Østhagen 2020; Østhagen 2018b). Thus far, the other Spitsbergen Treaty signatories have accepted this, though Russia, Iceland and (at times) the EU have been outspokenly critical of what they perceive as Norwegian Coast Guard discrimination against their fishing vessels (Pedersen 2009, 2017; Molenaar 2012; Østhagen 2018b; Østhagen and Raspotnik 2018). Note that the current dispute concerning the status of the maritime zone around Svalbard will not be dealt with here because it is not a boundary dispute per se, although it will be touched on in boundary discussions concerning Denmark/Greenland and Russia.[2]

7.1   UNITED KINGDOM

A key development occurred in 1964, when the UK informed Norway that it wished to start negotiations based on the equidistance principle (Ø. Jensen 2014a, 66). The UK wanted an agreement with Norway before dealing with other, more complicated boundary issues further south involving Denmark, Germany, the Netherlands, Belgium and France (Ryggvik 2014). Offering to use the equidistance principle was a concession because the UK ratified the Geneva Convention that same year that looked to favour using the shelf as the gauge. Norway's response to the British offer was immediate and positive.

The Norwegian side was also pleased at the willingness of the British negotiators to accept a boundary calculated from straight baselines drawn between the outer islands and reefs along Norway's highly fragmented west coast (Kindingstad 2002). The UK had previously challenged those straight baselines before the ICJ, which had ruled in Norway's favour in the 1951 Anglo-Norwegian Fisheries Case (Green 1952). That being said, the UK bene-fitted from the fact that the Shetland and Orkney Islands were likewise granted full effect regarding a calculation of the equidistance line.

The agreement between Norway and the UK was concluded in 1965, just one year after negotiations began (Norwegian Petroleum Directorate 2017). Jensen (2014a, 66) argues that it was not only prospects of oil and gas that drove the British and Norwegian interest in maritime delineation in the North Sea, but also clarity around sedentary species due to the importance of fisheries in the area. Any knowledge about oil and gas in the area was, in any event, limited at the time.

7.2   DENMARK

Negotiations with Denmark proved more difficult. Denmark had ratified the Geneva Convention in 1963 and could have been expected to argue that Norway's continental shelf was bounded by the Norwegian Trench (McHarg et al. 2010), the deepest part of which lies between Norway and Denmark. However, Denmark had a strong interest in seeing the equidistance principle applied in the south so that it could define its boundary with West Germany. The latter held that the location of the boundary should not be based on a simple application of the equidistance principle but should take into account the length of the coastline (Oude Elferink 2013).

Denmark would have also been aware that an argument based on coastal length was equally available to Norway because the length of the Norwegian coast facing Denmark greatly exceeds the length of the Danish coast facing Norway. Accepting the application of the equidistance principle with Norway enabled Denmark to be consistent in its legal arguments and avoid the worst-case scenario of having to make concessions based on coastal length in both the south and north. Norway and Denmark concluded their boundary agreement in 1965 (Oude Elferink 2013). Denmark was also interested in a quick settlement of the boundary with Norway so that oil exploration in the northern portion of its North Sea continental shelf could begin (Ryggvik 2014).

The Norway–Denmark boundary agreement was a win–win result for both countries. Denmark was able to secure a straightforward application of the equidistance principle in the north before being forced to accept qualifications to that principle in the south. Norway avoided any challenge to its position that might have been based on the Geneva Convention, gaining jurisdiction over a portion of the North Sea equal in size to its entire land mass (Ryggvik 2014).

Having agreed to a straightforward application of the equidistance principle in 1965, Norway and Denmark had no difficultly agreeing to do so again when, in 1979, they settled the very short maritime boundary between Norway and the Faroe Islands, which involved a relatively straightforward application of the principle (Kindingstad 2002).

Another boundary dispute emerged in 1980 when Denmark extended its 200 n.m. fisheries zone northwards along Greenland's east coast, creating an overlap with the Norwegian zone on the northwest side of Jan Mayen (Churchill 2001). Jan Mayen is a small island located roughly 250 n.m. east of Greenland and 360 n.m. northeast of Iceland. It has belonged to Norway since 1930, when Norway claimed sovereignty over it through historic title. There is no permanent population on Jan Mayen (apart from a few scientists on rotation from the Norwegian Meteorological Institute), but the EEZ around the island supports a sizeable fishery.

Denmark argued that it deserved a larger proportion of this disputed zone because the coast of Greenland is much longer than that of Jan Mayen and because the population of Greenland, living much closer to the area, deserved privileged access to the fish stocks located there (Churchill 1994). Norway, in contrast to its Jan Mayen dispute with Iceland, held firmly to the equidistance principle. After years of unsuccessful negotiations, Denmark submitted the dispute to the ICJ in 1988 (Churchill 1994).

In this Norway–Denmark dispute, the ICJ delimited a single maritime boundary between Greenland and Jan Mayen in 1993 (ICJ 1993). The court began with an equidistance line on a provisional basis and then considered whether 'special circumstances' justified any adjustments to achieve an 'equitable result'. The court concluded that the longer length of the Greenland coast required a delimitation that tracked closer to Jan Mayen and that the line should also be shifted slightly eastwards to allow Denmark equitable access to fish stocks. Norway and Denmark implemented the judgement through a boundary treaty concluded in 1995 (Hoel 2014, 55; Churchill 1994). The interests of fisheries played a role in the dispute, though these were mostly on the side of Norway's negotiating partners. To some degree, this was recognised in the ICJ judgement, which adjusted the Norway–Denmark boundary to accommodate Greenland's interest in a potential capelin fishery (Churchill 1994, 3–6).

Finally, Norway and Denmark turned to the Svalbard archipelago in the early 2000s. As previously outlined, Norway's sovereignty over the Svalbard archipelago was recognised by the 1920 Spitsbergen Treaty. To avoid escalating the dispute with other countries over the scope of the treaty and possible rights of access to offshore oil and gas resources, Norway has not claimed an EEZ around Svalbard (Hønneland 2013). Instead, in 1977, Norway claimed a 200 n.m. FPZ around Svalbard for conservation purposes, arguing that this zone is not covered by the treaty because such zones did not exist in maritime law in 1920 (Pedersen and Henriksen 2009; Churchill and Ulfstein 2010). However, under international law, a state does not need to claim a continental shelf because that is automatically generated by the adjoining territory (McDorman 2009, 21–34). Norway claims that Svalbard does not have a continental shelf in its own right (legally speaking; of course it has one in the physical sense) and that the continental shelf around Svalbard, as an extension of the mainland's continental shelf of the Norwegian mainland, is solely under Norwegian jurisdiction.

Norway drew straight baselines around Svalbard in 2001; Denmark drew straight baselines around Greenland in 2004 (Anderson 2009, 373–84). Then, in 2006, Norway and Denmark concluded an all-purpose maritime boundary between Svalbard and Greenland (Denmark–Norway 2006). Roughly 430 n.m. long, the boundary is based on an equidistance line, adjusted slightly to take into account the presence of Denmark's Tobias Island some 38 n.m. off the Greenland coast (Oude Elferink 2007).

By concluding the treaty, Denmark implicitly recognised that Svalbard generates both fishing and continental shelf rights although not taking a position on the question of whether these are covered by the Spitsbergen Treaty. The boundary treaty includes a provision on straddling mineral deposits, whereby either party can initiate negotiations on possible cooperative solutions without committing both parties to any result. The preamble of the Svalbard–Greenland Delimitation Agreement also specifies that the treaty does not set the boundary between their respective extended continental shelves that stretch northwards towards the North Pole, which is a matter that the parties will need to address sometime in the future (Thomassen 2013, 30).

Economic interests seem to have provided some motivation for the Norway–Denmark negotiations. Oude Elferink (2007, 376) explains how the 2006 treaty's provisions on straddling mineral deposits are based on the 1995 treaty regarding the boundary between Jan Mayen and Greenland, with further details on how exploitation would occur. The inclusion of these detailed provisions anticipates oil and gas activity along the new boundary at some point. For Norway, an additional goal was the international recognition for its position that Svalbard is indeed entitled to a fishing zone and continental shelf, even though the status of the waters and seabed around Svalbard have yet to be settled.

7.3   ICELAND

In June 1979, Iceland adopted a 200 n.m. EEZ, just as Norway had done along the coast of its mainland three years earlier (R. R. Churchill 2001, 118). The new Icelandic zone came within 200 n.m. of Jan Mayen, so Norway responded by declaring its own 200 n.m. maritime zone (Fisheries Zone) around the island, creating an overlap (Churchill 2001, 118). Norway then took the view – consistent with its approach to other maritime boundaries at the time – that the equidistance principle was an appropriate solution. In contrast, Iceland held that it should have a greater proportion of the disputed zone given that the rights of the two states in this instance were generated by a small, remote, uninhabited island, on the one hand, and a significantly larger, populated island nation, on the other hand (Ryggvik 2014).

The dispute between Norway and Iceland was resolved through a conciliation committee consisting of three members: one from Norway, one from Iceland, and one from the USA, as a neutral third party (Linderfalk 2016). An agreement was signed in 1981, whereby the Icelandic continental shelf was recognised as extending a full 200 n.m. from the Icelandic coast in the area between Jan Mayen and Iceland, notwithstanding the proximity of the Norwegian island (Norway–Iceland 1981; Ø. Jensen 2014a, 72–73).

Thus, Iceland gained a much larger continental shelf than it would have had under the equidistance principle. At the same time, a resource-sharing regime was incorporated into the new boundary agreement. Norway gained the right to participate in 25% of the oil and gas exploration on a portion of Iceland's continental shelf just south of the new boundary, while Iceland gained the right to participate in 25% of the oil and gas exploration on a portion of Jan Mayen's continental shelf just north of the new boundary (Norway–Iceland 1981).

Norway's willingness to concede to Iceland's position was based on several political and economic considerations. First, insisting on the equidistance principle in the context of a small, remote and unpopulated island would have damaged the relations between Norway and its smaller Nordic neighbour (Ryggvik 2014). Second, Norway had already discovered large oil fields in the North Sea, whereas Iceland had no equivalent resources (Kindingstad 2002). Third, the most promising oil and gas prospects between Iceland and Jan

Mayen were located close to the latter, in an area that Norway had received, despite its concession (Linderfalk 2016; Ryggvik 2014).

The dispute has also been connected to larger considerations regarding NATO membership and anti-NATO sentiment in Iceland at the time (Jóhannesson 2013). Thus, we cannot discount the broader security context given the Cold War and Iceland's role in the Greenland–Iceland–UK (GIUK) gap vis-à-vis the USSR,[3] though this does not seem to have been the primary impetus for the agreement.

In 2008, as oil prices peaked and prospects of actual oil and gas activity came into view, Norway and Iceland concluded a follow-up treaty that provided a more detailed framework for cooperative exploration of straddling deposits and deposits within the two zones, here following a rule of 25% participation (Iceland–Norway 2008). According to then Norwegian Foreign Minister Jonas Gahr Støre, the arrangement provided the predictability that the oil companies needed (Karlsbakk 2008).[4] This joint hydrocarbon regime, although not unprecedented (McDorman 2009, 333–37), was the first to be established in Arctic waters. Regardless of these developments, the area around Jan Mayen is relatively inhospitable to petroleum development, with difficult ice conditions and deep water (Churchill 1994, 6), and no large-scale drilling has since commenced.

7.4   RUSSIA

A dispute arose in the 1960s when Norway and the Soviet Union both drew on the 1958 Geneva Convention to claim continental shelf rights (United Nations 1958; Henriksen and Ulfstein 2011). The dispute then escalated in 1977 when both countries had asserted 200 n.m. EEZs encompassing both fish and seabed resources (Henriksen and Ulfstein 2011). An agreement concerning the disputed area, the 'Grey Zone Agreement', was signed in 1978 and was renewed annually up until 2010 (Riste 2005, 250). It recognised the rights of Norway and of the USSR/Russia to fisheries in the area, without tackling the boundary dispute as such (Stabrun 2009).

For more than four decades, Oslo and Moscow contested roughly 175,000 km², or about 10%, of the Barents Sea. Leaning on the 'sector' principle, that is, the use of meridians to determine a boundary instead of equidistance, Moscow argued that various 'special circumstances' were relevant to the boundary delimitation: the length and shape of Russia's coast; the size of the respective populations in the adjacent areas; ice conditions; fishing, shipping and other economic interests; and strategic concerns (Østreng and Prydz 2007; Moe, Fjærtoft, and Øverland 2011; Hoel 2014).

Norway responded that the Soviet Union had drawn the line in 1926 for the sole purpose of defining the territorial status of several offshore islands, without any intention of delimiting maritime zones. It argued that a median line should instead be drawn from the mouth of the Varangerfjord, a narrow inlet between Finnmark and the Kola Peninsula, within which a territorial sea boundary had been agreed upon in 1957 (Churchill and Ulfstein 1992, 47). Such a line would be equidistant at all points from the Norwegian and Soviet mainland coasts; further out, it would be equidistant from Svalbard in the west and Novaya Zemlya and Franz Josef Land in the east (Churchill and Ulfstein 1992, 63).

After being informally initiated in Oslo in 1970 (Ø. Jensen 2014a, 75) and formally launched in 1974 (Moe, Fjærtoft, and Øverland 2011, 47), negotiations over the Barents Sea boundary continued for almost four decades. In 2002, Russian President Putin visited Oslo and agreed with the Norwegian Prime Minister Bondevik that boundary dispute negotiations should be continued to find a solution quickly (Ø. Jensen 2014a, 82). Three years later, they announced that their countries would initiate 'strategic cooperation' on petroleum development in the Barents Sea (Bakken and Aanensen 2010). Negotiations on the boundary dispute were resumed later that year, when a new government had taken office in Oslo.

Norwegian Foreign Minister Jonas Gahr Støre placed the Arctic on the top of the political agenda in 2005–2006 (Støre 2012; Østhagen, Sharp, and Hilde 2018; Østhagen 2021a). This included, in particular, the Barents Sea dispute and Norway's relations with Russia. In 2007, the two countries signed a revision of the 1957 agreement on the boundary within the Varangerfjord (Russia–Norway 2007). This revision provided a clear starting point for the boundary farther out and was an essential step towards full resolution of the dispute (Norwegian Government 2007).

The breakthrough on the remaining boundary issues came in 2010, when the two countries committed to an all-purpose boundary that would be drawn 'on the basis of international law in order to achieve an equitable solution', recognising 'relevant factors ... including the effect of major disparities in respective coastal lengths' while dividing 'the overall disputed area in two parts of approximately the same size' (Norwegian Government 2010). The resulting treaty, with geodetic lines connecting eight defined points, was ratified by the Norwegian and Russian governments after the Norwegian Storting and Russian Duma gave their consent in 2011 (Gibbs 2010; T. Neumann 2010; United Nations 2010).

The 2010 treaty sets a single maritime boundary, delineating both the EEZ and continental shelf within 200 n.m. from shore and the extended continental shelf beyond that. It is only of limited interest as to 'whether the agreed boundary is best described as a modified median line (as argued by Norway) or a modified sector line (as argued by Russia)' (Henriksen and Ulfstein 2011, 7) because the treaty divides the previously disputed sector almost in half. It also includes provisions on comanagement of any hydrocarbons that straddle the boundary through the conclusion of a 'unitisation agreement' for the exploitation of any such deposits and on the access of private companies to drilling rights on either side of the boundary (Byers 2013, 43–44; Fjærtoft et al. 2018).

That the dispute could finally be settled was because of several factors, not least the potential for oil and gas (Moe, Fjærtoft, and Øverland 2011; Hoel 2014; Claes and Moe 2018). Although the prospects for oil and gas were an important driver in getting the boundary issue onto the bilateral political agenda, the presence of resources probably also made the topic more sensitive. 'The negotiations would have been easier if there was no suspicion of oil and gas resources in the area', the lead negotiator on the Norwegian side told a Norwegian newspaper in 2005 (Dagens Næringsliv 2005; Blomqvist 2006). Additionally, the negotiator highlighted energy and fisheries as the key points in the negotiations. Yet as Blomqvist (2006, 60) shows through a discourse analysis concerning the boundary before it was settled, 'It seems that until the debate over petroleum increased, the unresolved boundary was at large seen as unproblematic.' Oil and gas seem to have had multiple effects here, contributing to the push to find a solution while also making it challenging for the negotiators.

In addition to oil and gas, fisheries have long been at the forefront of the cooperative maritime relationship between Norway and Russia (Hønneland 2012). The Barents Sea is home to the world's largest cod fishery (Stokke, Anderson, and Mirovitskaya 1999; Stokke 2000; Hønneland 2013). Over the last decade, effective management cooperation has enabled Norway and Russia to increase their science-based quotas – to the point where the cod stock provides more than 2 USD billion in sustainable annual catches (Hønneland and Jørgensen 2015). However, the fisheries did not serve as an incentive for concluding the boundary treaty in 2010. As explained by Geir Hønneland (2013; see also Solstad 2012), some Russian fishermen voiced concerns – both before and after – that a clear delineation would deny access to some historically important fishing grounds. Also, in Norway, there were concerns over losing access, in tandem with domestic pressure to ensure that quotas remained at preagreement levels (Solstad 2012, 78–82).

Moe, Fjærtoft and Øverland (2011), as well as Holsbø (2011) and Solstad (2012), hold that beyond economic interests, Russia's desire to affirm the primacy of the LOS regime and also 'tidy up its spatial fringes' help to explain the 2010 settlement. Indeed, Russia has benefitted greatly from the right of every state to an EEZ because of its extremely long coastline. In addition, the shallow nature of the Arctic Ocean means that Russia will also benefit from the LOS rules on extended continental shelves, perhaps more than any other country. Eliminating the legal and political uncertainties associated with unresolved maritime boundary disputes is one way of securing these benefits vis-à-vis third parties that might challenge Russian positions in the Arctic (Holsbø 2011; Moe, Fjærtoft, and Øverland 2011).

7.5   LOOKING TO RUSSIA AND THE ARCTIC

From the 1960s onwards, successive Norwegian governments have maintained a policy of actively seeking to resolve maritime boundary disputes. This policy has emerged as the result of several factors. The first, identified by Oxman (1995, 254) regarding boundaries worldwide, is 'the desire to "consolidate" coastal state jurisdiction newly acquired under international law'. In the North Sea, Norway sought rapid settlements with the UK and Denmark after the Geneva Convention, and parallel developments in state practice made it possible to present credible claims for a 200 n.m. continental shelf (and EEZ) (Ryggvik 2014).

Norway was also thinking strategically – beyond the North Sea to its contested Barents Sea boundary with the Soviet Union. As Norway's position in the Barents Sea was based on equidistance, any new state practice in favour of that principle in the North Sea could be seen as bolstering its claim in the High North. Regardless, a more general desire to consolidate rights was apparent in the Barents Sea, where economic and security interests motivated the negotiation of a clearly defined boundary with the Soviet Union and later Russia (Moe, Fjærtoft, and Øverland 2011, 147). The adoption of the 2010 Barents Sea Boundary Treaty came as the result of more than four decades of continuous effort by Norwegian diplomats.

Thus, economic interests have long been a factor in Norway's efforts to resolve its boundary disputes. The negotiations with the UK and Denmark came in relation to the possibility of substantial hydrocarbon reserves in the North Sea. The motivation provided by economic interests was powerful enough to overcome concerns about incomplete knowledge regarding exactly where those resources were located. Although this uncertainty loomed large in the negotiations (Ryggvik 2014), the influx of interested foreign companies and the prospect of win–win outcomes carried the negotiations forward.

In the post-Cold War era, renewed interest in Arctic affairs has also played a role, especially in the resolution of the Barents Sea boundary dispute. Norway's Arctic policy should be seen as the result of economic incentives aligning with broader foreign policy goals: safeguarding Norwegian sovereignty and ensuring stability in regional relations (I. B. Neumann and Gstöhl 2006; I. B. Neumann et al. 2008; Haraldstad 2014; Østhagen 2021a). The new focus on the Arctic was coupled with Norway's long-standing policy of pragmatic cooperation with Russia on transboundary issues ranging from fish stocks to migration and trade (Hønneland 1999; Østhagen 2016). Proactively settling maritime boundaries became more than a technical, legal or economic issue for Norway: it emerged as a core element of Norwegian foreign policy.

Finally, it is noteworthy that Norway was willing to depart from the equidistance principle in negotiating certain boundaries while maintaining its commitment to the principle more generally (Byers and Østhagen 2017, 54). The Jan Mayen–Iceland boundary illustrates this, with concessions being made in light of Iceland's dependence on fisheries and Norway's positive disposition towards its smaller Nordic neighbour (Oxman 1995, 259; Churchill 1994). However, when Denmark raised similar arguments concerning the Jan Mayen–Greenland boundary, Norway remained unrelenting until the ICJ delimited the boundary in 1993.

These were arguably calculated moves that allowed Norway to settle individual disputes amicably while preserving its general negotiating position in favour of equidistance – not least in the Barents Sea. At the same time, Norway used resource-sharing/joint development regimes in the Iceland–Jan Mayen, Greenland–Svalbard and Barents Sea boundary treaties. These arrangements differed regarding the details but were all aimed at overcoming the uncertainty barrier – the unwillingness of states to settle boundaries because of concerns that they might be surrendering access to still undiscovered seabed resources.

NOTES

1.   The Swedish–Norwegian maritime boundary that was determined in 1909 through international arbitration is not dealt with here. See the Permanent Court of Arbitration 1909 for more information. The boundary is limited in legal, political and economic importance, and settlement took place outside the temporal scope of this book (1960–current date; see also Ø. Jensen 2014, 62–63).
2.   For more on this dispute, see Pedersen 2009 and Østhagen and Raspotnik 2018.
3.   The maritime space in the Northeast Atlantic between the UK, Greenlandic and Iceland was important for denial of USSR access to the Atlantic during the Cold War.
4.   According to the same report, the new treaty was signed just three days after the Bank of Norway granted the Icelandic government a loan of approximately €1 million as part of Norway's assistance to Iceland during the global financial crisis.

8.   Legal context and precedent

Having examined four different countries and their range of maritime boundaries (settled as well as unsettled), we can now turn to the various mechanisms
that help explain why some boundaries remain in dispute, while others have
been agreed on. Central here are three overarching branches of logic: (1)
legal context and precedent, (2) the interaction effect between oil and gas and
domestic interests, and (3) the link between security concerns and fisheries.
The following three chapters will explore these in depth.

Starting with the legal: As maritime boundaries are a modern invention of
international law, which is itself constantly (if usually slowly) evolving, legal
factors are central to understanding the causal mechanisms behind dispute
settlement. We may surmise that the legal factors – the effects of international
law and the legal aspects of how disputes stand vis-à-vis each other – are
highly influential in driving and, in other instances, obstructing the settlement
of maritime boundary disputes between states (see Chapter 2). Which legal
factors, however, will depend on the context of the boundary itself regarding
international law and the other boundaries of the disputing states, as well as
the origin of the boundary in question. At the same time, by utilising legal
factors such as origin or precedent in explaining political outcomes, we can
advance our understanding of these factors in relation to those outlined in the
other sections.

8.1   LEGAL ORIGINS

First, an examination of the origins of a boundary dispute can provide information about how willing a state is to yield on its position. Most of the maritime
boundaries examined here originated with the implementation of the 200 n.m.
maritime zones from the 1960s until the mid-1980s. As noted, this innovation
in international law provided the main impetus for the emergence of maritime
boundary disputes in the first place. In many cases, the ensuing disputes over
where to delineate the exact boundary between opposing or adjacent coastlines
became a simple matter of negotiations driven by a legal rationale – LOS Art.
74 and Art. 83 – that was employed when the zones were first established.[1]

Consequently, we need to consider the timing of maritime boundary agreements as well and distinguish between two groups of maritime boundaries: (1)
those that were settled/negotiated as a result of developments in international

law in the general period from 1960 to around 1985 and (2) those that remained disputed for various reasons and may or may not have been settled in the period from around 1985 to 2020.

To some extent, the former group of boundaries can be explained in conjunction with developments in international law. This we can see in the cases of Australia and Indonesia/Solomon Islands/France; Colombia and Ecuador/Panama/Costa Rica; Canada and Denmark/France; as well as Norway and Denmark/UK/Iceland. International law was not the sole rationale for agreeing on a maritime boundary in these instances, but it was arguably the main impetus and, in some cases, also a driver for settlement. As in the case of Norway–UK (1965) or Canada–Denmark (1973), the desire to secure a clearly demarcated maritime boundary rapidly so as to enable oil and gas exploration and/or expanded fisheries weighed more than uncertainty over the exact location of the same resources.

In such instances, the driver of settlement was a combination of an interest in resources (as discussed above) and the desire to implement extended maritime zones quickly before possible challenges to the new rules either internationally or to the exact boundary by regional third parties developed. This latter point was especially prominent in the North Sea maritime boundary disputes between Norway, the UK, Denmark, the Netherlands, Belgium and West Germany (Oude Elferink 2013) and for Colombia in its disputes with the Dominican Republic, Haiti, Jamaica, Honduras and Costa Rica (Londoño 2015). Thus, the emerging need to delineate space – through the implementation of 200 n.m. zones – made settlement a relatively simple affair as the precedent of extended maritime zones was solidifying.

However, in those cases where the implementation of extended maritime zones amplified an already underlying disagreement as to where to delineate between adjoining territorial seas, a dispute arose, in which compromise was difficult. We see this in the case of Canada–USA in the Beaufort Sea, where the dispute stemmed from a diverging interpretation of the 1825 treaty between Russia and Great Britain; in the case of Canada–USA in the Dixon Entrance where the two states disagreed over the meaning of an arbitration decision in 1903; and in the case of Colombia–Venezuela over the status of the Gulf of Venezuela, where the land border had been settled in 1941 but not the maritime boundary. In both the Beaufort Sea and Dixon Entrance, Canada's legal position hinges on what can be called 'hard points' (Byers and Østhagen 2017, 60): the treaty concluded between Britain and Russia in 1825 and the A–B line drawn by an arbitration tribunal in 1903.

The same can be said about the Colombia–Nicaragua dispute. Although there was not initially an underlying disagreement (as the spatial extent of the maritime zones had been agreed in 1928), the maturation of international law and domestic politics combined helped cast the pre-LOS agreement from

1928 in a negative light. This eventually led Nicaragua to reject that treaty, in turn resulting in three decades of Colombian efforts to secure maritime claims against Nicaragua through state practice with its other neighbours.[2]

However, in other cases where the origin was not fixed to a specific historic decision or treaty but arose because of differing legal interpretations or practices at the time of the 200 n.m. expansion, more flexibility was afforded in negotiations. In the case of Norway–Russia, the Varangerfjord Agreement was signed in 1957 concerning territorial waters between the two countries, and the divergence in positions on the EEZ boundary was due to each state favouring a different legal principle in its expansion. Eventually, compromise became possible as equidistance gradually gained international acceptance through ICJ rulings and state practice. This weakened Russia's adherence to the sector principle, and Russia could relinquish its stance, now able to gain around half of the disputed area without reneging on its legal position or earlier treaties (Moe, Fjærtoft, and Øverland 2011). Further, Russia stood to gain enormously from submissions for an extended continental shelf under Art. 76 in the LOS because adherence to the LOS in the Arctic would be a beneficial strategy.

Similar conclusions can be drawn from the case of Australia–East Timor, where the boundary positions were not grounded in previous treaties or historical decisions but instead awarded the states enough flexibility to alter the conditions for the boundary and the joint zone from 1999 onwards. The same goes for Canada's negotiations with Denmark over Greenland, which, in contrast to many of its boundaries with the USA, were not based on a colonial period agreements; this also holds true for Norway's dispute with Denmark over Jan Mayen, where Norway refused to relent on the equidistance principle, despite the differences between the opposing coasts, instead agreeing to send the case to the ICJ and abide by its verdict.

In sum, the boundaries agreed in the 1960s–1980s have generally involved applications of the equidistance delimitation method – or a variant thereof – because states found themselves in the fortunate position of being able to claim extended maritime zones. Thus, there was an inherent win–win situation in the negotiations that followed at that time, a point I return to in Chapter 11. Disputes over those boundaries that remained without agreement or became settled only after the legal regime had solidified in the 1990s–2010s seem to be largely the result of diverging interpretations of delimitation methods, treaties or historical decisions that originated before the concept of extended maritime zones came into being.

Clearly, then, the legal origins of a boundary dispute have considerable influence on the flexibility afforded to states when negotiating where to demarcate a given boundary. However, we must again distinguish between what acts as a barrier to settlement (e.g., treaty origins) and what acts as a driver (e.g., desire to consolidate the LOS or affirm boundaries regionally vis-à-vis other states). This latter point is the basis of the second legal point discussed below. We can argue that not having boundary positions bound to historic agreements predating the LOS is not necessary to achieve settlement, although it certainly seems to be an advantage.

8.2   PRECEDENCE AT SEA

Second, a crucial legal component of any settlement or dispute is legal precedent. This is a concept that differs in nature between international law and national law (e.g., Byers 2000; Lang and Beattie 2009; De Brabandere 2016). Although precedence does not hold the same role in international law as it does in various national legal systems, states are concerned about – and sensitive to – how their actions and agreements align vis-à-vis customary international law and to their other outstanding disputes, where relevant (e.g., Cohen 2015; Brabandere 2016).

When a state has its eyes set on larger legal–political objectives in conjunction with developments in international law, it may be willing to forgo some of the underlying uncertainty inherent in any maritime boundary compromise as a way to reap the benefits of securing larger gains in terms of a finalised boundary that seems the best outcome available. This is also an argument for viewing a country's maritime boundaries and disputes as a collection to comprehend how that country views each dispute in relation to the others and whether this has any impact on negotiations.

We see this across the states studied here. Perhaps the clearest example of how legal motivation spurs a state into willingness to compromise is Colombia. Noting its significant and complex land and sea dispute with Nicaragua over the 1928 Esguerra–Bárcenas Treaty, Colombia sought to bolster its sovereignty over both the San Andrés Archipelago and its maritime zones by rapidly settling all its maritime boundaries with nearby states, except that with Nicaragua. Similarly, to bolster it claims vis-à-vis Venezuela in the dispute over the Gulf of Venezuela, it sought settlements with Haiti and the Dominican Republic.

Further, we can note a similar concern for legal precedent (related to the tendency to regard maritime boundaries as related to each other) in Norway, Canada and Australia. Australia developed its whole northern maritime frontier in relation to other states (Indonesia, East Timor, Papua New Guinea). Canada still fears that abandoning its stance in one of its unsettled maritime boundary disputes with the USA might impact its position in its other remaining disputes. Norway made explicit efforts to safeguard its stance concerning the use of equidistance, with the Barents Sea dispute with Russia in mind. The lead Norwegian negotiator for the maritime boundary with Russia in 2010 argues that the compromise was built around mechanisms concerning resource sharing, which were based on the previously revised agreement with Iceland in 2008. This, in turn, was built on decades of cooperation with the UK on resource sharing in the North Sea (Norwegian Diplomat I 2017).

It is apparent that Australia, Canada, Colombia and Norway have all considered how their boundary agreements stand vis-à-vis both international legal precedent and the status of their own outstanding boundaries. In some cases, like that of Australia–Solomon Islands (1988), Australia–New Zealand (2004), Norway–UK (1965) or Norway–Denmark (1965), the desire to set a legal precedent, whether regionally or regarding another specific outstanding maritime dispute, acted as a motivating factor for settlement in that specific case. The same can be said of the special case of Svalbard, where Norway has been attempting to build a legal case for its position through the boundary agreement with Denmark–Greenland (2006). As explained by Oxman (1995, 265):

> Others [states] may wish to use one or more agreements to influence an outstanding delimitation either directly or indirectly. The classic example of this approach is the equidistant line drawn by Denmark and the Netherlands as part of a more general implementation of the equidistance principle in Article 6 of the Convention on the Continental Shelf in the North Sea that included, in addition to these two states, Norway and the United Kingdom. It represented not only an attempt to reinforce the use of equidistance in the North Sea but, by extending the line to a point equidistant from their coasts and the German coast, an effort to apply equidistance directly to their respective boundaries with Germany.

The clearest example of the driving power of this factor is the case of Colombia's Caribbean 'labyrinth' (Londoño 2015). The desire to build regional legal practice concerning the zone around the San Andrés Archipelago (and surrounding islands/cays/reefs) regarding the dispute with Nicaragua, as well as the dispute with Venezuela in the Gulf of Venezuela, spurred Colombia into settlements with Panama, Costa Rica, the Dominican Republic, Haiti, Jamaica and Honduras. However, the desire to bolster a particular method of delimitation does not only have a positive or enabling effect for maritime boundaries. In some instances, Colombia's strategy backfired – as with Honduras (due to domestic opposition and Nicaraguan pressure) and Costa Rica (where the Pacific 1984 agreement had to be decoupled from the 1977 Caribbean agreement to avoid domestic opposition).

A similar concern for legal precedent is found in the case of Canada–USA (Juan de Fuca) and in the Norway–Denmark ICJ case, where one or both parties proved unwilling to settle for fear of establishing a legal precedent that would be unfavourable in other remaining disputes. More generally, many of Canada's unresolved maritime boundary disputes seem related to concerns about legal consistency and the creation of precedents, in addition to their historic origins. In the Gulf of Maine case, Canada was concerned that advancing an equidistance-based argument would weaken its position in the Beaufort Sea and Dixon Entrance disputes. Therefore, it reframed the argument to focus on equity considerations – which, not coincidentally, led to an equidistant result (McRae 1989, 155). The dispute seaward of Juan de Fuca Strait touches on the legality of Canada's straight baselines, which is also a central issue in the Canada–USA dispute over the status of the Northwest Passage (Byers and Lalonde 2009; Pharand 2007). Canada might worry that a compromise seaward of the Juan de Fuca Strait would weaken its position in the Arctic.

These examples show how having multiple boundary disputes with the USA has posed a sequencing problem for Canada, in which resolving any given dispute almost always requires concessions from both sides. In 1977, Canada sought to solve the sequencing problem by offering to negotiate a 'package' deal – an offer refused by the USA that probably thought that dealing with each boundary dispute in turn would work to its overall advantage. Norway's sequencing problem has always concerned its dispute with Russia, which could be resolved only based on some negotiated version of 'equity'. Norway dealt with the problem by resolving its other boundaries first, thereby freeing it to offer concessions on equidistance during negotiations over the Barents Sea boundary with Russia.

In the case of Colombia–Nicaragua, however, Colombian efforts to undermine the Nicaraguan position and uphold the 1928 treaty by co-opting neighbouring states to support its position spurred resentment and opposition in Nicaragua, as well as among Columbia's domestic audience. Thus, Colombia's political efforts to set a positive legal precedent – which, in turn, spurred many of Colombia's own maritime boundary negotiations in the 1970s and 1980s – proved to be a major barrier to settlement with Nicaragua.

However, it is interesting to note that with legal practice developing and the politics surrounding maritime space also adapting, some of those quickly agreed boundaries might appear less favourable than at the time of agreement. The cases studied here offer many examples of this. In the case of Norway and Denmark (1965), Danish politicians of the time have been criticised for giving away oil and gas resources (Kaarsted 1992). In the case of Australia and Indonesia (1971–73/1981), the Indonesian foreign minister spoke of his country as 'being taken to the cleaners' in connection with the seabed agreements of the early 1970s (Kaye 2001, 21). In the case of Colombia and Honduras/Costa Rica, Colombia's eagerness to finalise boundaries in the late 1970s to reap the benefits of recent developments in international law and bolster its case vis-à-vis Nicaragua led to resentment afterwards when the political fallout became apparent. In the case of Timor-Leste, the agreed boundary arrangements had been heralded as a fair compromise but were later – in light of resource and political developments – deemed so unfavourable to Timor-Leste that it withdrew from the whole arrangement and forced Australia to compromise in another agreement (2018) through an LOS conciliation process. In turn, there have been discussions about the possibility that the agreement with Timor-Leste could cause Indonesia to question its agreements with Australia in the same area.

Moreover, states try to influence regional customary international law through negotiated settlements, as seen in the North Sea and Caribbean Sea cases. How successful this strategy has been, however, can be questioned. In the case of Colombia, there was a strong belief that four decades of developing a regional legal precedent regarding the dispute with Nicaragua would lead to a favourable ICJ verdict. (Whether the outcome was in fact favourable, however, is a different debate.) Still, regional customary international law is not that straightforward (Cohen 2015). As explained in Chapter 2, customary international law does not simply derive from state actions or judicial decisions by international courts. Denmark experienced this in the North Sea cases with West Germany, where its bilateral agreements with the UK and Norway on equidistance boundaries did not end up swaying the ICJ to accept regional precedents in favour of equidistance.

Likewise, we see signs – as also theorised by Huth, Croco and Appel (2011) and Allee and Hurth (2006) – that states turn to international courts to find a suitable outcome when domestic opposition to concessions is too great. This could be said to have been the case in Norway's boundary arrangements with Iceland (1981) and Denmark (1995); in Canada's arrangements with the USA over the Gulf of Maine (1984) and with France (1992); and in Colombia's case with Nicaragua. A crucial point that arises from these cases is the extent to which states have concern for international law. If an ICJ decision or international arbitration proves unfavourable to states in maritime boundary disputes, will states accept that ruling?

In the 1984 Gulf of Maine ICJ case, the 1985 Iceland–Norway arbitration and the 1993 Norway–Denmark ICJ case, the parties chose to adhere to the outcome of the adjudication/arbitration. Australia, however, has chosen to withdraw maritime disputes over maritime zones from the terms in its acceptance of the ICJ and LOS, even though it did engage with Timor-Leste under LOS conciliation in 2017. More extreme, in the 2012 Colombia–Nicaragua case, Colombia chose to ignore the ruling and even withdrew from further proceedings in the court because of distrust and discontent with the outcome. In another case worth noting, the Philippines initiated arbitration proceedings in 2013 under Annex VII to the LOS against China's 'nine-dash line' in the South China Sea. China declared it would not participate in the arbitration nor adhere to its rulings. On 12 July 2016, the arbitration tribunal ruled in favour of the Philippines, but the ruling was rejected by China (PCA 2016).

A relevant question here is whether we are witnessing increasing distrust of international arbitration and adjudication when it comes to maritime boundary disputes. Such a trend might go in tandem with growing public awareness about maritime issues and outstanding maritime disputes (Kleinsteiber 2013; Nyman and Tiller 2020). As also put by Oude Elferink, Henriksen and Busch (2018a, 399), 'The analysis and conclusions of some of the chapters of this volume [on maritime boundaries] do suggest that normativity in the delimitation process extends beyond the basic rule as contained in Articles 74 and 83 of the LOSC [Law of the Sea Convention] as reflected in customary international law'. Individual states might indeed refer to this 'normativity' when dissatisfied with the outcome of international processes or as a bargaining strategy for future negotiations (Goldstein et al. 2000, 396), as noted in the case of Australia–Timor-Leste and Colombia–Nicaragua. The alternative is simply to keep the dispute unsettled, as with several cases between Canada and the USA, which have been characterised by unwillingness to utilise third-party dispute resolution mechanisms.

To conclude, there cannot be any doubt that international law sets the parameters for maritime boundaries and related dispute resolution. What is sometimes not specified, however, is how these effects are evident across cases – a dimension often ignored in much of the political literature on disputes, boundaries and the maritime domain.[3] To grasp the importance and effect of the legal regime for the oceans, we must study how the LOS sets the parameters for bilateral state negotiations (e.g., Nemeth et al. 2014). Even when the dispute is kept out of international courts and tribunals, contemporary legal precedent has been shown to impact bilateral negotiations because states lean on legal arguments in the reasoning behind their claims (Prescott and Schofield 2005; Allee and Huth 2006; Ásgeirsdóttir and Steinwand 2016).

We have also seen that states conceive of their maritime boundaries as a set of boundaries, not as single data points. The countries examined here – especially Canada, Colombia and Norway – exhibit a strong desire to bolster the legal precedents of their approach when settling maritime boundaries because of some other more complicated or sensitive boundaries in need of settlement. Further, we have seen that states are not only concerned with, or aware of, current precedence – they also actively attempt to shape it through their bilateral settlement proceedings. However, questions remain as to the effect of these efforts on customary international law, as well as exactly how far states are willing to yield to international law when faced with an unfavourable ruling.

NOTES

1. It may be discussed whether 'dispute' is in and of itself an accurate description of these boundary issues because the states negotiated and settled rather quickly when the zonal overlaps emerged.

2.   This point also concerns the dispute over the maritime zones around Svalbard – not directly dealt with in this book – which hinges on the Spitsbergen Treaty from 1920, a time when the concept of extended maritime zones around the Archipelago had not been developed.

3.   Simplified, the dilemma is that IR is concerned with *why* questions, sometimes neglecting the role of law, whereas international law is focused on the *content* of law itself, sometimes ignoring its political context (Snow 1913; Byers 2000; Hafner-Burton, Victor, and Lupu 2012).

9.   Oil and gas and public perception

Let us now turn to what is often described as the most significant factor for the resolution of a maritime boundary dispute, at least historically: the presence of oil and gas resources. This has often figured as the primary argument for why states seek settlement of boundary disputes at sea. However, it is not always clear how the presence of hydrocarbons influences the willingness of states to make concessions or not. Sometimes the presence of fixed shelf resources can instead cause dispute, even conflict between parties. We thus need to add another dimension to further nuance this factor, namely the role of the domestic audience and public perception. This chapter explores these mechanisms further.

9.1   TO DRILL OR NOT TO DRILL

Does economic activity and special interests in a maritime domain lead to a greater likelihood of settlement, or does it hinder going to a settlement because states are eager to maximise their potential gains? Access to resources is frequently used as an argument for why states engage in settling disputes, despite the uncertainty as to exactly what causal mechanisms are involved. As I have noted, there is a variation in the extent to which economic interests in hydrocarbon resources play an active part in prompting boundary settlement. Three different patterns can be identified here.

In the North Sea disputes between Norway–UK/Denmark in the mid-1960s, hydrocarbon resources were an obvious factor in driving negotiations forward. The same applies to the negotiations between Australia and Indonesia in the early 1970s, the failed Beaufort Sea negotiations between Canada and the USA around 2010, the case of Australia–East Timor 2000–2018 and the cases of Norway with Russia (2010) and Iceland (1981).

In some other cases, oil and gas resources and related interests were involved but were not the major factor. This applies to Canada and the USA over the Gulf of Maine, Canada with France over St. Pierre and Miquelon, Canada with Denmark over Greenland in both 1973 and 2012, Norway's boundary agreements with Denmark/Greenland over both Jan Mayen (1995) and Svalbard (2006) and Colombia's agreements with Honduras (1986) and Jamaica (1993).

In the final category of cases where oil/gas was a relevant factor, the presence of resources acted as a barrier to reaching an agreement, unlike the two categories above. This concerns in particular the case of Papua New Guinea, where oil and gas licences had already been awarded by Australia and, thus, had to be dealt with in the negotiations. In several other boundary agreements, such as Norway–Russia (2010), Canada–USA (Beaufort Sea) and Colombia's ongoing disputes with both Nicaragua and Venezuela, oil and gas have played a negative role in barring final agreement, despite also spurring the countries onwards.

Important here, however, is not the fact that there are resources located in a given area but that there is a marked interest in resource development. For such resources to hold considerable relevance, there must be special interests revolving around this resource. These may range from oil and gas companies themselves to regional or local businesses or governance structures. For example, there might be oil and gas resources located on the continental shelf between Australia and New Zealand in the Tasman Sea, where a boundary was drawn in 2004. Limited knowledge of the seabed in this part of the world, combined with extreme remoteness, depths of more than 5000 metres and limited infrastructure, made potential oil and gas resources almost irrelevant in the boundary negotiations. And in the case of Canada–USA in the Beaufort Sea, negotiations on the boundary were initiated after oil prices rose in the 2000s but were suspended when prices fell.

Therefore, on the one hand, we would assume that the presence or expectations of hydrocarbons would make states less willing to compromise, attempting instead to claim as much as possible of the disputed area. Take the US–Canada Beaufort Sea case: boundary negotiations were initiated in 2010, but uncertainty concerning the existence and location of hydrocarbons seems to have contributed to the suspension of the talks. An effort was made to resolve the uncertainty through seismic mapping of the disputed zone, but the resulting delay coincided with a sharp drop in world oil prices.

In fact, the absence of economic interests may facilitate an agreement, as Oxman (1995, 251) notes, concerning US success in settling maritime boundary disputes far from home: 'The most obvious explanation is that it is easiest to reach agreement in the case of small islands surrounded by the deep waters of the Caribbean Sea or the Pacific Ocean where the boundary regions are unlikely to contain hydrocarbons or localized fisheries.' In Canada, this same factor may have contributed to the conclusion of the tentative agreement in the Lincoln Sea with Denmark/Greenland, where the area in dispute was small and the prospects of economic activity very low.

On the other hand, a clear boundary is usually needed if the state is to reap the benefits of the potential resources in a disputed zone, which could be assumed to prompt states to settle more readily. Compare the above-noted case of Canada with that of Norway, which was willing to concede a large area of contested seabed to Iceland in the 1980s because it knew that the greatest potential for oil and gas lay close to Jan Mayen. In these instances, the presence of resources plays a part in explaining why a settlement was obtained.

As put by Renouf (1988, iii), 'The resolution of offshore boundary disputes must first be attained before managed development of the offshore region under dispute may begin'. Governments do not directly engage in oil and gas exploitation themselves: they benefit through private or state-owned companies and the economic benefits that these companies produce. Companies need stable operating environments and will be reluctant to invest in exploration and production if there is uncertainty as to which country has sovereign rights over which resources. This is particularly relevant for the oil and gas industry, given the extremely costly investments and sunk costs in licences and infrastructure in contrast to, say, fisheries.

This speaks to a crucial component of the early stage maritime boundary delimitations and international law and international politics more broadly: the potential positive-sum dimension of delimiting maritime zones. From having only 12 n.m. (in many instances even smaller) maritime zones, states suddenly acquired expansive zones with the rights to marine living and sedentary resources in the 1960s–1980s. In some instances, ensuring that these zones became established took primacy over limited disputes over baselines and equidistance. As put by one scholar:

> The view that prevailed in the Caribbean during the latter seventies was that delimitation itself constituted the first step towards economic development. Thus, it was regarded as more of a casual stimulant than a consequent response to current or potential economic activities. (Nweihed 1993e, 523)

Some of this development occurred a few decades later as well, when states were getting ready to submit claims to the CLCS for extended continental shelves because of the 10-year deadline after each of their individual ratifications of the LOS. Australia and New Zealand agreed on a continental shelf boundary in 2004 in advance of their submissions in the hope that this would help ensure their rights. Similarly, the Norway/Denmark agreement on the seabed and EEZ boundary between Greenland and Svalbard in 2006 would seem to be connected to these countries' submissions to the CLCS.

Therefore, uncertainty is not an absolute barrier to a boundary agreement. In the North Sea in the 1960s, Norway, Denmark and the UK decided that the cost of leaving boundaries unresolved was higher than any potential losses resulting from uncertainty as to whom would obtain the greatest share of the seabed resources. If, in some cases, states were to agree on maritime boundaries because of an economic interest in the disputed area, whereas in other cases they could do so because of limited economic interests, which mechanism would hold true and when?

Most states (i.e., their leaders and diplomats) negotiate with an eye towards the future and are inherently risk adverse when giving away space and rights that might prove valuable in the long term.[1] A key difference, however, between, for example, Norway and Canada in the aforementioned cases has been the willingness of Norway to use hydrocarbon cooperation regimes as a way of reaching final settlements. Although there is a provision on hydrocarbon sharing in the 1973 Canada–Greenland boundary treaty, it does not commit the parties to any procedures or outcomes. In addition, whereas the 2012 tentative agreement between Canada and Denmark on the Lincoln Sea foresees the inclusion of rules on hydrocarbon cooperation, that part of the treaty has yet to be finalised (Global Affairs Canada 2018). In contrast, Norway has hydrocarbon-sharing mechanisms built into most of its boundary treaties, including, most significantly, in the Barents Sea with Russia.

A key measure employed to alleviate concerns over exactly where resources are located on the seabed – uncertainty – is the use of resource-sharing agreements concerning straddling deposits. However, that mechanism cannot solve the issue if the dispute concerns a relatively large area because both parties would still have an interest in moving the boundary to their advantage before agreeing to a 'straddling deposits' agreement. One way of circumventing this problem was applied in the case of Norway–Russia (in 2010): any resources developed within the whole formerly disputed area would be shared equally. This can mean a relatively large concession for both parties if they staunchly believe that their own claim is more legitimate. This makes the demand for a maritime boundary itself less immediate, perhaps even dispelling the need for it altogether, though there are more reasons than only function (resources) for a boundary at sea.

Thus, we can note positive-sum outcomes to instilling cooperation across maritime boundaries. By settling maritime disputes, states can engage in previously risky resource development and possible joint extraction. Clear jurisdictions as to whom owns what (or, more accurately, which rights) enable states to cooperate. Although the loss of space is in and of itself a zero-sum game because of the nature of delineation (if one state acquires more of a given maritime territory, the other necessarily loses the same amount), joint management of transboundary resources that enable further development of these areas constitute positive-sum situations with mutual gains (VanderZwaag 2010; Ásgeirsdóttir and Steinwand 2016).

Finally, I note a shift between the boundaries settled in the 1960s–1980s and those settled later, some of which are still disputed. From being able to accept the inherent uncertainty pertaining to the location of seabed resources and settling a boundary regardless – at times with an eye towards larger political or legal aspirations – states seem to have become less willing to forgo potential economic gains by granting concessions in a boundary negotiation. Only when highly advanced resource-sharing agreements have been developed to overcome the uncertainties have states been willing to accept an agreement – sometimes not even then.

Interestingly, the feat of turning a formerly disputed maritime area into a shared regime makes the maritime boundary itself less relevant, for oil and gas extraction at least. In some of the cases outlined here – notably Colombia–Jamaica (1993) and Australia–Timor-Leste (2018) – a joint zone or regime replaced the need for a boundary. Although in the latter case, dissatisfaction with that arrangement eventually prompted a regular maritime boundary between both parties. Others have argued for similar zones to be developed to avoid conflict over the exact location of a maritime boundary, most noticeably in the dispute between Guinea-Bissau and Senegal (Okafor-Yarwood 2015) and in the South China Sea (Beckman et al. 2013).

That being said, states do care about the actual location of the 'invisible' boundary at sea.[2] As put by a Norwegian diplomat (2017) after the Russia negotiations, 'It is inherently difficult with a condominium. We need a proper border in place in case of future problems arising, and then we can develop solutions within the formerly disputed area. This helps reduce conflict'.

9.2   THE VETO PLAYERS

It is obvious that the presence of hydrocarbon resources in a maritime area plays a central role in explaining both why the area itself attracts political attention and why states are willing to concede spatial claims in favour of a demarcated domain. However, we cannot equate the presence of hydro-carbons with the settlement of a boundary. First and foremost, there must be interest in developing the resources in question. Otherwise, the resources themselves will have no effect. Moreover, it is not sufficient to explain the interests themselves and whether they are active in a given dispute: we must examine the political system and power structure in each of the countries.

In this book, there is a clear divide between Australia and Canada, on the one hand, as federated countries where the states/provinces have an active say in matters of domestic and foreign policy, and Colombia and Norway, on the other hand, as relatively centralised and/or unitary states where the departments/counties are less autonomous or active in (inter)national politics.

Therefore, it is not surprising that in all the questions concerning Norway's maritime boundaries, from the 1960s up until 2010, the relevant county governments in the south and the north are absent in terms of ratification procedures and/or vetoing potential agreements. In addition, although the Saami people have significant rights under Norwegian law, none of those rights extend beyond the territorial sea (Ravna 2012). The Norwegian Parliament (the Storting) is required to ratify any agreement made by the government, but because governments derive from a parliamentary majority or can act only if supported by the majority, it would take extraordinary circumstances for a boundary agreement not to be ratified. The only issue here arose with Russia in 2010, when Russian regional opposition to the agreement almost toppled it in the Russian Parliament (Duma) (Hønneland 2016).

The same set-up concerning the national legislature applies to Australia and Canada, where the governments are based on a parliamentary majority, although regional governments also have a say. As to the cases at hand, the most blatant example concerns the region of Queensland and how it was involved in the process of negotiations between Australia and Papua New Guinea in the mid-1970s. Queensland acted as both a proponent and hurdle because the regional government had strong interests in preserving the economic interests of its northern inhabitants (Burmester 1982). In that boundary dispute, the Australian Parliament also played an active role in deliberating the legality of the settlement with reference to the constitution, as Australia was 'giving away' its northernmost islands to Papua New Guinea (Willheim 1989; Kaye 1994). Park (1993b) argues that elections taking place in both countries in 1977, which reaffirmed voter confidence in the sitting government in Canberra and gave the first democratic mandate to the Pangu Party in Papua New Guinea, helped the governments agree on a boundary.

Canada has several maritime boundary disputes that are complicated by provincial claims and perhaps even constitutionally entrenched rights. It is difficult to imagine the governments of British Columbia and New Brunswick standing by quietly while the national government of Canada negotiates with the USA over the Dixon Entrance or Machias Seal Island. Similarly, the 'Inuvialuit Final Agreement' might represent a complication for Canada in the Beaufort Sea boundary dispute.

In Canadian–US negotiations over the Gulf of Maine before the submission to the ICJ in 1981, it became clear that the US Senate and the Canadian provinces of Nova Scotia and New Brunswick were opposed to a compromise that would mean giving away maritime space (and resources). With the US Senate, formal approval was needed for ratification of a signed agreement between the two countries. Thus, the countries agreed to send the dispute to the ICJ. For Canada, the same engagement of regional veto players was evident in the Machias Seal Island dispute, in the dispute with France over St. Pierre and Miquelon, in the Juan de Fuca dispute and in the Dixon Entrance dispute (Byers and Østhagen 2018).

In the latter two cases, the provincial government of British Columbia has been highly active, even issuing a statement of its own position on the disputes (British Columbia 1977). The Machias Seal Island dispute involved the same actors as those in the Gulf of Maine dispute. In addition, in the dispute with France over St. Pierre and Miquelon, the province of Newfoundland had strong interests. This stands in contrast to the disputes with Denmark/Greenland (1973/2012), where the regional interests of the territory of Nunavut were less engaged in the disputes, also because of the sheer distance from local communities to the maritime boundaries in question, as well as the fact that Nunavut only came into existence in 1999. The difference must be understood as relating to the difference in regional autonomy awarded to Canada's three northern territories in contrast to the relatively autonomous southern provinces. The Beaufort Sea dispute stands out. It is not the regional government (the territory of Yukon) that is directly involved, though it holds interests. In the dispute, the Inuvialuit Settlement Region, as an indigenous governance structure, has a considerable say on the final rejection or acceptance of any agreement, alongside the US state of Alaska.

In Colombia, the executive branch (the president) and the parliament are separated in the typical presidential-style system (e.g., Lijphart 1989). Thus, despite limited regional autonomy in terms of maritime boundaries and dispute agreements, the Colombian Parliament could pose a challenge to the agreements made by a sitting president. This was evident in some of Colombia's maritime boundary agreements, though that has been in connection with general public dissatisfaction with the agreements (see below), not system-specific processes like those seen in Australia and Canada.

In sum, it is the structure of the state itself – the domestic institutional set-up – that determines a country's ability to get a maritime boundary agreement with a third country ratified and in force. As per Putnam's (1988) two-level games, state leaders must balance 'veto players' against their preferred outcome in international negotiations (Tsebelis 1995). In particular, regional governments may have considerable influence in these processes, if the political system allows for it. In systems where regional governments have less autonomy or influence, there is considerably less flexibility to thwart or influence boundary arrangements.

In turn, this domestic institutional factor is predominantly a barrier to settlement. In none of the instances examined in this book did the systemic set-up and ratification procedures prompt states to pursue maritime boundary agreements. This is yet another factor that state leaders must bear in mind and, in some instances, overcome to reach an agreement.

9.3   PUBLIC PERCEPTION

Next, we must see the function of these resources in conjunction with other factors that expand beyond institutional procedures and veto players. Domestic opinion concerned with 'giving away' rights and the active involvement of fishers afraid of losing their livelihoods due to 'big oil' interests may constitute significant challenges that the existence of potential resources may not be sufficient to overcome. States and their leaders are highly sensitive to public opinion and political pressure. Naturally, this varies from country to country and context to context. An authoritarian regime might be less influenced by public opinion concerning its maritime boundary agreements. However, even autocrats are sensitive to public pressure and may be wary of causing too much dissatisfaction with a political outcome (see Wiegand 2011b; Hensel et al. 2008).

This is also why negotiations are often kept secret when states embark on them, here for fear of arousing public engagement and, thus, side-tracking further negotiations before a final deal can be made. We have seen this in the case of Norway–Russia (2010), the case of Canada–USA over the Beaufort Sea, and the case of Australia–Timor-Leste (2018). Although state leaders and/ or lead negotiators may confirm to the media or through official channels that negotiations are underway or taking place,[3] the point is to avoid having explicit details of the agreement leaked before a final agreement has been achieved.

Public engagement and/or pressure can interfere later in the ratification processes. Politicians have no way of avoiding the formalised set-up, which can act as a barrier to a settlement. Whereas those processes are structured, public engagement in the matter is informal. Politicians and state leaders may utilise this engagement to their advantage, stirring up positive or negative sentiments to reinforce certain positions. Strong popular engagement might influence the chance of ratification through official structures, or regional governments' active participation in boundary processes may spur further popular engagement.

In the boundary disputes examined in this book, the public engagement factor has been evident to varying degrees. In the case of Australia–Papua New Guinea (1978), the (re)election of governments already engaged in boundary-making helped prod the process towards its finalisation. Another straightforward impact of popular engagement was seen in the case of Australia–East Timor, with the Timorese gaining independence in 1999 and ending with a new agreement in 2018. Public outcry and domestic pressure in both countries, also relating to the espionage revelations in 2012, led the Timorese government to negate the previous agreements and start ICJ proceedings against Australia. By 2017, domestic opinion in both countries had succeeded in spurring the Australian government to renegotiate, demanding a fairer agreement for Timor-Leste.

Canada has also experienced the pressures of popular engagement in its boundary negotiations. There was a spillover effect from the power of Canada's provinces and local structures (in the case of the Beaufort Sea) to thwart eventual agreements and their ability to foster public support for their positions. Further, there was public opposition to settlement in the Gulf of

Maine, Dixon Entrance and St. Pierre and Miquelon cases. Here, two interrelated issues entered the picture.

First, the economic interests of the affected provinces/regions helped spur negative public engagement, marked by fears of losing access to local resources (fisheries). Second, general popular opposition to agreements – between Canada and the USA in this instance – tie into regional patterns of amity and enmity and identity politics with another state (Kirkey 1995, 56). The political sensitivity of Canadians to the power differential with the USA should not be underestimated. Many of the great political debates of Canadian history have involved proposals to connect Canada more closely to its southern neighbour – whether through trade and investment agreements, improved access for US cultural industries or closer military cooperation (Brooks 2010, 379). This complicated history of how Canada and the USA relate and have interacted over decades also shapes the reactions of the general public to possible boundary agreements. For Canadian politicians, compromising with the USA – or being seen as giving in to US pressure – is unpalatable, and this factor helps explain several of the outstanding maritime boundary disputes with the USA. Thus, it is arguably not so much the actual power asymmetry as the perceptions of this among the general public (and thereby national politicians as well) that enter the picture. In comparison, Canada's boundaries with Denmark/Greenland, agreed upon in 1973 and 2012, generated limited public interest and could be concluded rather easily.

Similar factors have been prominent in Colombia's disputes. Whereas its first settled maritime boundary with Ecuador in 1975 was criticised for being irrelevant – a sign of the limited public interest in maritime domain at the time – negative conceptions of the 'other' in the maritime boundary negotiations with Nicaragua and Venezuela led to heavy public opposition to a compromise over the San Andrés Archipelago and Gulf of Venezuela, respectively. These sentiments, as noted in Chapter 6, seem to be strong in both Colombia and the opposing states, not allowing much leeway in negotiations. State leaders in all these countries have used the ongoing disputes in their domestic political campaigns to foster public support, which is a sign that there has been a considerable change in public engagement in maritime boundary disputes since the boundary with Ecuador was criticised for being 'inconsequential' four decades earlier.

Indeed, these boundary disputes have become so politicised that Colombia's efforts to secure related boundary agreements have encountered another form of domestic popular opposition due to the potentially harmful effects of the agreements on relations with Nicaragua (Londoño 2018). This has been evident especially in the cases of Colombia–Honduras (1986) and Colombia–Costa Rica (1977/1984). Domestic audiences can set limitations for what their governments can achieve in negotiations or how easily the results can be ratified. Similarly, Russian leaders faced strong domestic opposition in 2010 after the agreement with Norway, as well as regional attempts to block the agreement, though not enough to stop the process.

Finally, we should not discount the effect of domestic policy pushes or efforts that, in turn, influence the settlement of maritime boundary disputes. In the case of Canada–Denmark (tentative agreement 2012), the dispute was part of a broader Arctic policy 'push' on the part of the Harper government that aimed at settling outstanding disputes. We can note similar dynamics in Colombia's maritime disputes, particularly with Honduras in 1986, when the government instituted an active policy towards the Caribbean, which led to a focus on bolstering maritime claims and settling disputes. In Norway, the government's Arctic focus starting in 2005 prompted a desire to settle disputes with both Denmark/Greenland (2006) and Russia (2010), which reaped acclaim from the Norwegian public (L. C. Jensen 2017; Østhagen 2021a).

In sum, although national ratification procedures and veto players can be an effective barrier to the final settlement of a maritime boundary, larger popular engagement in maritime boundary disputes can alter an already agreed boundary settlement, engage politicians in seeking a settlement (as a way of gaining favour amongst the domestic audience) or dissuade them from doing so at all. The lines between national ratification procedures, popular awareness and engagement at large and economic interest are not crystal clear. However, it is important to understand why, in some instances, state leaders make use of maritime boundaries for political election purposes, despite limited economic or regional interest in the issue and why, in other instances, settling a boundary may prove impossible due to strong domestic opposition. There is an interplay between leaders and their domestic audiences when engaging in maritime boundary-making, which also explains why negotiations are often kept out of the public view.[4]

NOTES

1. Here, we can note the view, held by some, that Denmark was 'tricked' by Norway into conceding what become one of the world's largest offshore oil fields at the time. In 1992, the historian Kaarsted argued that the Danish foreign minister, Hækkerup, had conceded too much to the Norwegians in the negotiations (crucially, the considerable Ekofisk-oil field located just on the Norwegian side of the 1965 boundary) (Ø. Jensen 2014a, 68; Kaarsted 1992). However, that account seems to ignore the larger legal and political context of the negotiations, as well as the wider Danish interests in the North Sea.
2. 'We cannot see that border under water!' In a comical sketch, this was the answer to being asked whether it is punishable to fish – from a submarine – in Norwegian waters. 'Vi kan jo ikke se den grensen under vann!' Harald Heide Steen Jr. 'Ubåtkapteinen': https://www.youtube.com/watch?v=kaw3EqzxVbA.

3.   See, for example, analyses of statements made by negotiators in the Norway–
      Russia (2010) case (Blomqvist 2006; Solstad 2012).
4.   See Risse (2000) and Putnam (1988) for literature on international negotiations
      and the domestic audience.

10. Security and fisheries

Beyond the natural resources on the seabed and concern for legal precedent, a third and key dimension regarding why states settle their maritime boundary disputes when they do relates to the link between regional security relations and the risks of conflicts over fisheries. However, the role of fisheries is often overlooked when examining maritime boundary disputes, despite many states (and their negotiators) explicitly stating the desire to improve control and regulation of fisheries in the disputed zone(s).

Partly, this can be explained by the sometimes-contradictory effects of fishing interests: although authorities often favour enhanced regulation, the fishing fleets often do not actively seek this. Furthermore, we must link fishing interests with a central part of ocean geopolitics, namely security interests and the strategic and/or military role of the maritime domain in question. Only by establishing the link between power relations, strategic value, historic relations and fisheries are we able to tease out how both security and fisheries interests influence boundary making at sea.

10.1 THE VALUE OF MARITIME SPACE

Looking at the security of a state, a central point is the value of space itself, whether for the military (a certain piece of land might be strategically advantageous) or because of the general advantage of that territory to the state. Naturally, with the implementation of 200 n.m. zones, states' strategic interest in ocean space mushroomed, and it became increasingly important to protect sovereign rights in this newly acquired domain. However, with maritime space, the value itself – strategic or otherwise – becomes more challenging, especially given its physical nature and the limited possibilities for human presence at sea or for utilising ocean space itself beyond resource extraction or as an arena for encounters often related to non-ocean-based disputes.

Here, we must return to the difference between land and maritime space. The concept of occupation, which is pivotal in establishing title to land territory, holds no relevance in the maritime domain. This is a vital point because what we are discussing regarding states and maritime space are the sovereign rights to resources in the water column or on the seabed, not exclusive rights to the entire maritime 'territory' in question (beyond the territorial sea). We must bear in mind the crucial difference between sovereign rights (EEZ, continental shelf) and complete sovereignty as per Krasner's (1999) accounts. States cannot deny passage through their EEZs; they may only deny actors access to the marine resources there and apply environmental regulations within them.

Therefore, the strategic value of ocean space is predominantly conceived in terms of resources and the navigational interests of states. Yet as discussed, the main driver for settlement concerning resources does not seem to be the assumed value but interests in making use of relevant resources. That being said, awareness of a maritime area thought to hold considerable resources – even without the interest and/or capacities for exploiting those resources – can act as a major impediment to settlement. We must distinguish between the drivers for settlement and against it, where high strategic interests in terms of known or unknown resource value may hinder settlement, but – when acting as a driver – must be seen in conjunction with economic interests.

In some instances, the maritime area in question has indeed figured in larger strategic and/or security relations, as in the Colombia–Panama case related to the Panama Canal (1976), Canada–USA (Dixon Entrance – unresolved) related to US submarines, Norway–Russia (2010) related to the Barents Sea and the Russian bastion defence concept in that area (Olsen 2017) or Australia–Papua New Guinea (1978) related to the immigration and movement of people. However, all these contexts or relations have more concerned the specific bilateral or regional relationship between the actors involved than the strategic value of the maritime space in question.

Finally, the idea of the territorial conflict in and of itself has been relevant in some of the maritime boundaries examined. As scholars have shown (Holsti 1991; Vasquez and Valeriano 2009; Nemeth et al. 2014), the inclusion of territory in any interstate conflict makes resolution less likely and an armed conflict more likely. Governments often find it more difficult to give up (or risk giving up) land territory because land generally holds greater domestic political significance than the seabed or the ocean column (Hensel et al. 2008). It is worth noting that despite not figuring prominently in the cases examined (or indeed being the main focus of this book), the inclusion of territorial ownership or control in the disputes has had negative consequences for boundary delimitation. Inherently, maritime zones and their related boundaries derive from sovereignty over land, making this a crucial point for any potential settlement of maritime boundaries. As we have also seen in other maritime disputes, like that of Cyprus–Turkey–EU in 2019, if the land itself is disputed, that will render agreeing on a maritime boundary difficult, if not impossible. However, as seen in the Canada–Denmark case, there are sometimes ways of avoiding the territorial dispute and still establishing a full maritime boundary.

In sum, across the boundaries examined in this book, strategic value in terms of security and military factors has been less relevant than for territorial disputes on land. We must recognise the intrinsic difference between the rights that states have in extended maritime zones and the territorial sea (12 n.m.). Resources may make a specific maritime domain strategically important, even though 'value' on its own is not enough to enable a settlement. This may obstruct a settlement if uncertainty over where the resources are located or potential losses in a compromise trump the potential domestic gains of resource extraction. Moreover, the inclusion of a territorial dimension is a complicating factor.

10.2   POWER AT SEA

The strategic value of a given domain is only one way of assessing geopolitics at sea. Equally important are the broader power relations and notions of power asymmetry between actors at sea. A question arises regarding negotiations between states that inherently are based on law (see Chapters 2 and 8): Do asymmetrical relations confer ability on the superior state to coerce a favourable outcome? Or do they have the opposite effect: that the power disparity leads the inferior state to fear being coerced by the superior state, thereby making it less willing to compromise (see Fearon 1995)?

Several cases examined here have involved relations that were asymmetrical in terms of state-centred ideas of power.[1] This is relevant for Norway–Russia (settled 2010), Norway–UK (settled 1965), Australia–Papua New Guinea (settled 1978), Australia–Solomon Islands (settled 1988), Australia–Timor Leste (settled 2018), all of Canada's boundaries with the USA and, to varying degrees, all maritime boundaries of Colombia, except with Venezuela.

However, there does not appear to be strong supporting evidence for the effect of power asymmetry, whether positive or negative. No clear patterns emerge. The USA does not seem to have been able to utilise its favourable power disparity vis-à-vis Canada to coerce an outcome. Furthermore, Canada's unwillingness to settle with the USA (or US unwillingness to settle with Canada) cannot be fully explained by Canadian fears of being coerced because of the disparity in power (McDorman 2009). Notably, the willingness of Norway and Russia to settle did not derive from a change in power relations or from Russian efforts to coerce Norway (Byers and Østhagen 2017). According to several scholars, for decades, Norway had been more willing than Russia to compromise on the issue (Henriksen and Ulfstein 2011; Holsbø 2011; Moe, Fjærtoft, and Øverland 2011). The change came about because of factors like shifts in regional relations and domestic interests, not because of the balance of power between the actors.

In other words, the notion of power does seem to hold any direct effect when examining specific maritime boundary disputes. However, it does figure in the wider political context that any dispute finds itself within. In some of the cases involving Colombia or Australia – both having been the larger actor vis-à-vis most other countries in which they share a maritime boundary – there do seem to be hints of power disparity and its role. Concerning the boundaries with smaller states where Colombia sought to bolster its legal position vis-à-vis Nicaragua regarding the San Andrés Archipelago, Colombia actively pushed for an agreement as part of a larger regional strategy. In some instances, Colombia utilised naval superiority to enforce its sovereign rights: the boundaries were agreed despite the protests of local fishers from the opposing state. These fishing interests seem to have been less important for Colombia, being trumped by the importance of implementing boundaries. We see this in the cases of Jamaica (1993), Honduras (1986), the Dominican Republic (1978) and, to some extent, Costa Rica (1977/1984). In many instances, Colombia also ended up with rather favourable outcomes in terms of having the full weight to various cays and islands, as well as the recognition of the San Andrés Archipelago and its full effect in terms of maritime zones. Thus, we cannot rule out the possibility that power asymmetry did play a role as Colombia pursued settlement vigorously with its smaller Caribbean neighbours, even though the final outcomes of these boundary agreements were rather equal divisions of the disputed maritime space.

Similarly, in negotiating with Indonesia and Papua New Guinea in the early 1970s, Australia ended up with maritime and seabed boundaries that fully favoured its position and interests. Concessions can be seen in allowing Indonesian and Papua New Guinean fishers access to Australian waters – however, as pointed out by Prescott (1993b; 2002), Australian interests in these waters have been limited and the actual benefit for the other party can be questioned. Further, in negotiations with Timor-Leste, Australia seems to have stood firm for decades regarding what it prized most highly: gaining access and privileges concerning the Greater Sunrise hydrocarbon fields. However, in that case, as well as with Papua New Guinea, the extreme disparity between the countries in terms of power, as well as wealth, level of development and dependence on the oil and gas industry, seem to have had an opposite effect: Australian governments feared being seen as taking advantage of their superior position (Burmester 2019; Kaye 2001; Former Australian Government Official 2019).

Consequently, we cannot completely discount the influence of power relations in determining or influencing outcomes, though it does not seem to have had an obvious discernible effect. However, as seen in the cases of both Australia and Colombia, when power disparity is combined with an active superior state pursuing boundary settlement, the asymmetry in the relationship may serve as a way of fast-tracking the boundary negotiations, not necessarily due to the direct use of force or threats in negotiations but because of the superior state's ability to offer additional benefits, like side payments or issue linkage (see Wiegand 2011a, 5, 44), and pressure in other ways. If, for instance, the USA had decided to implement an active policy or strategy similar to that of Colombia or Australia in settling its maritime boundaries, Canada might have found it easier to achieve momentum in negotiations and might have been persuaded through various other means to accept a compromise. Although this does not mean that Canada is completely beholden to the interests of its larger neighbour, as the various boundary disputes outlined here between the USA and Canada have shown, Canadian unwillingness to yield – perhaps particularly vis-à-vis the USA – is not to be underestimated (McDorman 2009, 3).

Smaller and less influential states have more limited possibilities to engage in such an active push vis-à-vis its neighbours. When Norway refocused on its maritime boundaries in 2005 and 2006 in conjunction with its Arctic policy agenda, it found amenable partners in Reykjavik and Copenhagen/ Nuuk but was dependent on Moscow accepting the proposed compromise in the Barents Sea. In its special relationship with the USA, Canada experienced the same when, in 2006, Canadian Prime Minister Stephen Harper put Arctic sovereignty at the centre of his election strategy. However, Harper's political focus on the Arctic may have become a double-edged sword regarding dispute settlement because his strong rhetoric contributed to what has been called 'sovereignty anxiety' – the idea that Canada is struggling to uphold its sovereignty in the Arctic and, thus, is vulnerable to security threats in the region (see Griffiths, Huebert, and Lackenbauer 2011). This anxiety, in turn, would have made it politically more difficult to accept concessions as part of a boundary settlement, especially when the USA was the negotiating partner.[2]

For Canada in the Beaufort Sea and Norway in the Barents Sea, achieving a settlement was highly contingent on the preferences of the more powerful neighbour. The Barents Sea dispute was resolved when Russia became willing to make concessions, here motivated by a desire to achieve legal certainty regarding oil and gas and the development of its larger Arctic strategy. The USA did not show a comparable willingness to compromise because its economic interests were less involved and perhaps because of concerns that moving away from equidistance in the Beaufort Sea would weaken its legal position in Dixon Entrance, seaward of the Juan de Fuca Strait and elsewhere in the world (Byers and Østhagen 2017, 59).

However, the Norwegian and Canadian contexts are quite different. Norway sought to secure its sovereignty through the settlement of its boundaries – particularly with Russia – in a situation where the existence of a dispute posed security risks. By contrast, Canada's anxiety about its own sovereignty played the opposite role, acting as a barrier to settlement, even when managing ongoing disputes was a viable option because of the amicable nature of its relationship with the USA (Byers and Østhagen 2017; Østhagen, Sharp, and Hilde 2018). Only one Canada–USA boundary dispute has involved an explicit security dimension: the passage of US submarines through the Dixon Entrance. In this case, the two countries have essentially agreed to disagree, with Canada granting blanket permission for such passage and the USA insisting that permission is not required, while the issue itself is not of any real significance.

This is not to discount the potential effects and relevance of such efforts by weaker states in negotiations with a more powerful partner. However, it does show that the capacities and capabilities of states do matter for maritime boundary negotiations, albeit often indirectly.

10.3   PATTERNS OF ENMITY AND AMITY

In several boundary disputes, emphasis has been placed on 'good neighbourly relations' or 'strong bonds of friendship' in the wording of many of the agreements themselves – when settled – or in the statements surrounding them. However, we may question the relevance of such statements made when an agreement had already been achieved. Related notions of 'allies' and 'security relations'[3] could be indicating that, for instance, NATO members should have an easier time settling maritime boundary disputes than long-term rivals like non-NATO-member Russia and NATO-member Norway. However, I do not find any clear pattern concerning this when looking at the maritime issues examined. Norway and Russia did manage to agree on a boundary dispute that had dogged the bilateral security relationship for almost four decades. However, the two arguably closest allies in this study – Canada and the USA – have not agreed on a single maritime boundary through bilateral negotiations.[4] Australia and New Zealand managed to agree on their extended boundary, even though the maritime space in question was relatively uncontentious. Colombia, however, has struggled with Venezuela and Nicaragua because of regional security relations and legal reasoning and domestic interest groups.

A more interesting exercise involves attempting to uncover the historical patterns of interaction between the two actors and how these have fluctuated and shifted, at times creating room for 'friendly' relations and, at other times, deteriorating ones. Leaning on Buzan and Wæver, we can note that regional relations between actors – in this case, bilaterally – may compound over time, giving rise to patterns that might not make sense from a purely systemic point of view (Frazier and Stewart-Ingersoll 2010; Kelly 2007). That, in turn, has helped shape the identity of the relevant state and how it (its leaders and/or population) sees itself vis-à-vis others (e.g., Campbell 1998; I. B. Neumann 1996). Also important is the historical component involved because relationships may shift over time as various events impact relations (e.g., Hopf 2002).

For example, in the case of Australia–East Timor (2018), in the 1970s, Australia was already deeply engaged in the processes that led to Timorese independence and the eventual revisit of an already settled maritime arrangement. From having been less vocal against Indonesia's annexation of East

Timor in 1975 than some in Australia would have liked (Former Australian Government Official I 2019) to taking the lead of the UN force sent to prevent a civil war and facilitate the transition of the former colony into an independent state, relations deteriorated after 2006, with the Timorese now wanting a 'better deal' (Doherty 2017). Although regional relations can generally be said to have been positive between these two states, they turned sour regarding the maritime boundary dispute. In this case, the negative context acted as a hinderance to a settlement, which was not settled until the two countries took the step of making use of the LOS conciliation mechanism.

Another example is the Norway–Russia (2010) boundary. Although security relations bilaterally and in the Barents Sea specifically never truly 'warmed' after the end of the Cold War, the developments in Russia in the early 2000s and the decision of both countries to focus on Arctic cooperation (including joint oil and gas ventures) helped improve relations sufficiently to bring a compromise on the largest Arctic maritime boundary dispute within reach. That would hardly have been possible only a few years later, when relations deteriorated drastically with Russia's 2014 annexation of Crimea and engagement in the conflict in Ukraine, which was follows by the Norwegian response of imposing sanctions, in line with the EU and USA (Østhagen 2021a). A policy window where historic patterns of enmity were improved had opened but then shut almost as quickly. So here, regional patterns of enmity acted as a barrier whose removal could facilitate settlement, not as a driver of reaching a settlement itself.

Unsurprisingly, negative regional relations with the opposing state have impacted the willingness to settle an ongoing maritime boundary disputes. We see this in the above-mentioned cases, as well as with Colombia–Venezuela and Colombia–Nicaragua. We can question whether the possibility of settlement of the maritime dispute has been influenced by the larger negative relations or whether – as in the case of Colombia–Nicaragua – the maritime boundary is at the heart of the larger dispute itself, thus fuelling the negative relations further. The cases of Canadian–US boundaries and Norway–Denmark that took considerable time and even an ICJ ruling to settle show that the driver element of positive regional relations and historic patterns is limited. In these two instances, the countries are deemed close neighbours and culturally alike, yet achieving agreement on maritime boundaries has not been easy. Consequently, what drives settlement has less to do with historical patterns and regional relations and more with other factors, such as domestic interests and legal considerations.

Beyond a doubt, patterns of amity or enmity between disputing states do have an effect on the chance of states agreeing on a maritime boundary dispute. This holds true, to varying degrees of relevance and more so as a barrier than a driver of settlement. Regarding the latter point, one prerequisite is not to have a long history of negative relations between the actors involved. That, however, is not a sufficient condition for enabling a settlement. This is an important distinction, separating between regional relations as a barrier (with an obvious discernible effect across the cases examined) and as a driver (less of an effect). In other words, states do not settle maritime boundaries because they have close/amicable regional relations, but having a negative interaction makes it less likely that they will settle anything.

10.4   LINKING FISHERIES AND SECURITY CONCERNS

Finally, a central but often underappreciated dimension of maritime boundary disputes is fisheries and the role this economic activity plays with some of the security elements outlined already. In contrast to oil and gas extraction (see Chapter 9), the scale, history and sustainable nature of fisheries have made it an activity already existing in most maritime domains around the world and, thus, highly relevant regarding any extension of maritime zones and subsequent delineation of rights and access.

For example, in the case of Canada in the Gulf of Maine (1984) and around St. Pierre and Miquelon (1992), relatively high levels of fisheries and the potential for a 'cod war' scenario involving repeated and reciprocal arrests of fishing boats eventually pushed the disputing parties into adjudication and arbitration. Similar aspects have at times been relevant for Canada's unresolved disputes with the USA over Machias Seal Island, Juan de Fuca and the Dixon Entrance, where fisheries have acted as a hurdle to overcome.

We see the same fishing interests in almost all of Colombia's maritime boundaries in the Caribbean Sea; in Norway's boundary agreements with Denmark (Jan Mayen), Iceland and Russia; and in Australia's maritime boundaries with Papua New Guinea and Indonesia. In these cases, fishers in one or both disputing countries fear that a boundary will exclude them from important fishing grounds which, prior to the introduction of extended maritime zones, were open to all, without clear regulation. In contrast to the oil and gas industry, fishers have sometimes seen establishing a maritime boundary as a negative development.

Especially in the cases of Colombia's agreement with the Dominican Republic (1978) and Jamaica (1993), fishers from the two latter countries had to concede fishing grounds when the San Andrés Archipelago was given full effect in terms of maritime zones. This led to hostile encounters between the Colombian Navy and the fishers in those areas, though Colombia eventually gave concessions in the boundary agreements allowing for a joint zone/ access to fishers. Australia employed similar methods in its agreements with

Indonesia (1981) and Papua New Guinea (1978), allowing local/traditional fisheries from those countries to access newly delineated Australian waters.

The link between the specific interests of fisheries and their impact on political leaders should not be underestimated. In the case of Norway–Russia (2010), regional fishing interests in Murmansk almost obstructed ratification of the agreement in the Russian Duma in 2011 (Hønneland 2013).[5] In several of the other boundary disputes examined, fishing interests have blocked agreements or been a factor that had to be overcome in negotiations. The more important the industry for a specific region or country, the stronger the domestic opposition to 'giving away' maritime space and subsequent fishing rights.

In contrast to oil and gas interests, which have generally served as drivers for an agreement, the interests of the fisheries industry have tended to be a hindrance. However, for the healthy management of fish stocks, in particular those that are transboundary in nature, the extension of maritime zones and the clear delineation of boundaries have generally been positive developments. Here, we consequently have cases where fisheries have acted as a barrier to a settlement but have also – because of the need for clarification if fisheries and/or economic zones expand – sometimes prompted negotiation.

Hostile encounters over fisheries – mostly between one state and the fishers from another state – are frequent occurrences across the globe but often do not escalate further (Nyman 2013, 6).[6] Although the chances of a dispute occurring at sea are relatively high due to states' limited ability to exercise complete control over that domain, the chances of such a dispute escalating into a larger international conflict are fairly low (Tunsjø 2018, 128–130). It may be easier to limit disputes and clashes at sea when it comes to the wider ramifications because disputes concerning access to fisheries are usually seasonal (Salayo et al. 2006) and resolution is often possible due to their limited salience, tangible nature and the fact that they are divisible (Hensel et al. 2008). States have often managed to compartmentalise rather limited disputes over access to fisheries, often settling them (or not) without affecting their relations otherwise.

The Cod Wars and the Turbot War provide recent historical examples of rather severe conflict erupting over straddling fish stocks. In the case of the Cod Wars, access to fishing grounds was the initial cause for contention between the UK and Iceland. In the case of the Turbot War, excessive Spanish fishing in international waters just outside of the Canadian EEZ caused an ensuing conflict. These examples took place at a time when fisheries increased in magnitude and geographical scope, followed by an extension of the international legal regimes in the same domain (Swartz et al. 2010). Despite being rather prominent examples of conflicts involving fisheries, they did not escalate further.

When such incidents are taking place in the context of disputes over maritime space (in contrast to when states engage in a conflict in the maritime domain), the question is what strategic value the space itself holds. Limited encounters at sea often concern maritime space (fishing rights), but much depends on whether it is deemed expedient to escalate to achieve a strategic goal that might expand beyond the immediate dispute. Therefore, it is largely the context in which the fisheries incidents occur that determines the possible escalation of an incident.

Here, the idea of regional security complexes, or regional patterns of enmity and amity as explored previously, enters the picture. Proximity is important. The interactions (positive and negative) between geographically proximate states will be more intense and compound over time. This translates into regional security dilemmas – where the actions of another state are interpreted as offensive when they actually might be of a defensive nature, thus prompting defensive countermeasures that might be interpreted by the other state as offensive (Jervis 1978) – informed by shared histories (Buzan and Wæver 2003, 46). These ideas have been proven crucial for maritime boundary disputes across this book, which are inherently geographically and regionally bounded constructs between neighbouring states. In turn, the regional relations that 'compound' over time set the parameters for how states negotiate ongoing disputes.

It is in this context that fisheries often play a relevant role in the disputed maritime domain. Multiple dynamics are at play. On the one hand, fishers are not always interested in the clear delineation of maritime space that might lead to spatial exclusion. On the other hand, states and their officials are often interested in the delineation of the same space to uphold sovereign rights and hinder IUU fisheries, avoiding the uncertainty over who has authority and mandate. When this tension comes to the foreground in a disputed maritime domain, the conflict potential increases, thus also prompting further efforts by states to attempt to settle the boundary dispute to install clear jurisdictions. We have seen this explicitly across the cases examined: in Australia's boundaries with France, Solomon Islands, Papua New Guinea and Indonesia; in Canada's boundaries with the USA (Gulf of Maine in particular); in Colombia's boundaries with the Dominican Republic, Haiti, Honduras and Jamaica; and Norway's boundaries with Iceland, Denmark (Jan Mayen–Greenland) and Russia.

For example, fisheries played a considerable role in the Barents Sea dispute, initially as a barrier to settlement, especially on the Russian side. However, coupled with this were the underlying frictions and risks of a serious incident over fisheries and fishing access, which had been a constant fear in Norway from 1977 onwards (Holtsmark 2015).[7] The Elektron incident (see endnote) highlights the conflict potential inherent in state-to-state interactions concerning disputes over maritime space. In turn, these interactions within this specific maritime domain where the boundary dispute played a key part are crucial in explaining Norwegian sensitivities regarding Russian maritime actions in the region, but also why Norway and eventually Russia were keen to remove this thorn in their bilateral relations. Norway's relatively high tolerance for uncertainty regarding the existence and location of hydrocarbons when settling the 2010 Russia agreement can be explained partially by the counterbalancing desire to reduce uncertainties and risks of another kind: tensions and possible conflicts with a much more powerful country over competing claims to seabed resources in the Barents Sea.

One way of seeing it is that settling boundary disputes can reinforce sovereignty by removing sources of tension and potential conflict. This was Norway's view in the Barents Sea, where the 2010 agreement removed one source of tension and potential conflict with Russia. The annual grey zone agreement set out provisions for how to manage fisheries in the area without escalating incidents, though a more permanent framework was wanted by Norwegian authorities. This desire for risk reduction has seen Norway undertake ongoing efforts to 'tidy up its spatial fringes'[8] – especially regarding Russia, where it has been deemed in the national interest to settle a spatial dispute for fear of the inherent potential for escalation. Any conflict with Russia would necessarily threaten Norwegian sovereignty, given the power disparity between the two countries.

In Canada, however, where all of the boundaries are with NATO allies, there seems to have been greater tolerance for uncertainty over political relations with its neighbours, as manifested in the 'management' of disputes over maritime space. Canada–USA relations involve a similar power disparity as the Norway–Russia case, but this case is quite different. Canada and the USA are partners in NATO and the North American Aerospace Defense Command (NORAD) (Jockel and Sokolsky 2012; Zyla 2009). This greatly reduces the stakes involved in their boundary disputes and creates the sense that these disputes are 'manageable': there is no underlying security or political imperative for them to be resolved, at least not in the same manner. Also note that the issues regarding fisheries between the two countries have more or less been settled through bilateral mechanisms. As McDorman (2009, 195) argues, '[T]he allocation of government resources, both human and political, inevitably flows to the immediate and urgent' – even if it would be logical to resolve boundary dispute in the absence of 'immediate friction'.

In this respect, Canada is an outlier, being the country with the fewest maritime boundaries settled. All of these (if we consider the Lincoln Sea as tentatively settled) are with its southern neighbour. Fears of setting a legal precedent by concluding one dispute but not the others, as well as a domestic sentiment not inclined to accept settlements with the USA unless clear advantages are present (as perhaps possible in the Beaufort Sea in 2008–2010), has put Canada in a particularly difficult position. McDorman (2009, 3) refers to the 'emotional freight' of sovereignty disputes, especially for Canada via-à-vis

the USA. This 'freight' makes it politically more difficult to accept concessions as part of a boundary settlement with the USA. Hence, a power disparity
with its southern neighbour, combined with the fact that these disputes can be
'managed', deters the politicians in Ottawa from pursuing settlements with the
same vigour as their counterparts in Colombia, Australia and Norway.

Turning further south, in most of Colombia's maritime boundary disputes
in the Caribbean, fishing interests in the opposing state proved to be a significant hurdle for settlement. For instance, with Honduras, acquiescence to the
Colombian stance on the status of the San Andrés Archipelago meant giving
away access to fishing grounds of importance for local fishers. At the same
time, Colombia seems to have been mostly willing (except for Nicaragua and
Venezuela) to compromise on fishing zones to achieve larger legal–political
goals regarding the unresolved disputes or made use of joint zones where
fishing rights were shared through some bilateral mechanism.

The Australian approach with both Indonesia and Papua New Guinea and
later East Timor also speaks to the relevance of security interests coupled with
local fisheries and the recognition of the advantages of removing this potential
source of friction. Australia has had limited interest in fisheries in the maritime
domains bordering these countries to the north. For local fishers in the area,
however, these waters are essential in sustaining livelihoods, building on
centuries of traditional fishing that has traversed the recently introduced and
invisible maritime boundary. Consequently, in the agreements with Indonesia
(1981) and Papua New Guinea (1978), special provisions were made for
traditional fishers to access Australian waters. These arrangements have occasionally proven controversial, ranging from criticism of the Australian Navy's
heavy-handed rule enforcement to potential security challenges arising from
a fluid border arrangement. Nevertheless, the main underlying incentive was
the desire to remove sources of friction related to both security and local fishers
when the concept of a maritime boundary between the states was established.

In most cases where fisheries have been a considerable factor, this has
acted as a barrier to settlement because fishers have been concerned with the
loss of fishing grounds and being spatially restricted. However, as discussed,
when fisheries are coupled with a state's larger concerns over friction due to
an ongoing or unresolved maritime boundary dispute, the addition of security concerns can act as a strong incentive to pursue boundary agreement.
Therefore, a factor in explaining why states settle maritime boundaries is
the interplay between fisheries and security interests, in turn linking with the
concepts of regional security complexes and states' desire to implement clear
zones of jurisdiction.

Moreover, a central aspect of fisheries is the contrast with oil and gas
resources. Often lumped together as 'resources', these two types of maritime
economic activities have diverging effects on the process of settling maritime

boundary disputes. Whereas oil and gas conflicts generally provide strong incentives for settlement, because of oil and gas companies' limited tolerance for uncertainty over who owns the seabed where they operate, fishers seem to have a high tolerance for uncertainty due to the shifting nature of fisheries in general. Only, as shown here, when coupled with a country's desire to install clear jurisdiction and authority have fisheries acted as an impetus for boundary agreements.

With the effects of climate change on migratory fish stocks (Gänsbauer, Bechtold, and Wilfing 2016; Spijkers and Boonstra 2017; Stokke, Østhagen, and Raspotnik, n.d.), concerns over disputes arising over fish stocks and the need to delineate sovereign rights might further increase. The world's oceans are now being impacted in an unprecedented way, adding another layer to the challenge of international cooperation over fisheries. Wild fisheries are increasingly fully exploited, decreasing the total available biomass of marine resources (FAO 2016). At the same time, stocks are changing their migratory patterns because of changes in the geophysical marine environment (Allison et al. 2009; Brander 2010). These changing conditions are particularly troubling for international management of transboundary fish stocks, that is, fish stocks that move between and across neighbouring EEZs and high seas. Scholars foresee an increase in the failure of cooperation globally as the impact of climate change on fish stocks becomes increasingly apparent (Pinsky et al. 2018; Shearman and Smith 2007, 49–55; Cheung et al. 2017).

With almost half of all maritime boundaries worldwide still unsettled, numerous contexts exist where conflictual incidents at sea can occur between states and foreign fishing vessels, here because the clear jurisdiction of sovereign rights and inspection authority are not delineated.

NOTES

1. Note that power in this context refers to the traditional parameters often used, such as size (population, economy) and military capabilities (Mearsheimer 1995).
2. McDormar (2009, 3) refers to the 'emotional freight' of sovereignty disputes, especially for Canada via-à-vis the USA.
3. Adler and Barnett (1998) define a 'security community' in terms of the expectations of peaceful change among its members. This expectation comes about through a causal process. First, there are several 'precipitating conditions' that enable states to cooperate on security issues in the first place. These entail technological change, external threats and 'new interpretations' of social reality. As states (and the individuals representing these states) start to interact more frequently, this interaction transforms their 'possible roles and possible worlds'.
4. In terms of actual military cooperation, the two countries maintain close ties, with the Canadian military placing great importance on interoperability with US forces. The North American Aerospace Defence Command (NORAD),

established in 1957 to provide joint surveillance of potential air-space threats in North America, further testifies to the close security relationship between the two countries. NORAD has become one of the foundations of US–Canadian defence collaboration, with the surveillance of the maritime domain added in 2006 (Jockel and Sokolsky 2012). However, these arrangements are what Jockel and Sokolsky (2009) see as 'remarkably informal' within the broader context of North Atlantic security.

5.   Albeit beyond the scope of this book, in the case of Svalbard, starting in 2015, the problem was exacerbated when Norway and the EU became involved in a dispute over rights to licence snow crab fisheries, where specific industry interests in EU-member Latvia had hijacked the entire EU fisheries system (Østhagen and Raspotnik 2018).

6.   Examples include incidents between the Russian Border Guard (coast guard) and Japanese fishers around the Kuril Islands (Kaczynski 2007); between Vietnamese and Philippine fishers and the Chinese Coast Guard in the South China Sea (De Treglode and Buchanan 2016); between the Norwegian Coast Guard and Russian fishers in the Barents Sea around Svalbard (Østhagen 2018b); and between the US Coast Guard and Russian fishers in the Bering Sea in the 1990s (Conley, Melino, and Østhagen 2017). All these unfolded in the context of larger maritime disputes concerning access to fishing zones and/or maritime boundaries, with varying degrees of escalation related to the regional patterns of the relations between the states involved.

7.   In 2005, Norwegian fisheries inspectors from the coast guard boarded the Russian vessel *Elektron* in the fisheries protection zone around Svalbard, where Norway and Russia hold opposing views on the status of that zone. The captain of *Elektron*, in agreement with the Russian owners, decided to flee—with the Norwegian fisheries inspectors on board (Åtland and Ven Bruusgaard 2009, 339). From 16 until 19 October, four Norwegian coast guard vessels, as well as a maritime surveillance aircraft and several helicopters, tailed the trawler as it headed for Russian waters. The Norwegian Coast Guard considered boarding the vessel, but bad weather intervened (Åtland and Ven Bruusgaard 2009, 341). It is also highly likely that the Norwegian authorities were concerned with the escalation effects that an action like boarding the Russian vessel could have vis-à-vis Russia (Fermann and Inderberg 2015, 389, 395). On 20 October, the two fisheries inspectors were released by the Russian Border Service, which had arrived to escort *Elektron* to Murmansk after an 'intense dialogue' between the Russian and Norwegian governments (Skram 2017, 168).

8.   Paraphrasing Moe, Fjærtoft and Øverland 2011, 158.

11. The future of boundary disputes at sea

Having explored the different mechanisms that have led states to settle their disputes at sea, we can conceptualise a broader understanding of how changes at sea influence and determine the IR of ocean space, as well as state behaviour in the same domain. How has the growing focus on maritime space, as shown through this study of the various maritime boundary disputes, led to changes in the way states view and utilise the maritime domain for political purposes? Better understanding this can help comprehend not only why states settle their maritime boundary disputes, but also what the future brings for disputes over maritime space. Three trends deserve further consideration: (1) the continued institutionalisation of ocean space; (2) the growing symbolic relevance of the maritime domain; and (3) the increasing utilisation of oceans for various purposes that spark environmental concern.

11.1   THE DIFFERENCE BETWEEN LAND AND SEA

The oceans have generally been seen as the antithesis of land: opposite and inherently different – or as a void (Steinberg 2001). As already outlined, maritime space and its value for states have been defined as inherently functional. That may make it seem tempting to equate the value and importance of maritime space (and subsequent disputes over such space) to the functional value of that space. This appears to be the general trend in studies of territorially based conflict, being one of the rationales behind excluding maritime space in the first place. However, further nuancing is needed, not least concerning the distinction between land and sea. Precisely because it has qualities different from land, maritime space has been subject to extensive legalisation and a rights-based regime in favour of maritime states. A change began when all coastal states were awarded 200 n.m. EEZs. The expansion from 12 n.m. zones to 200 n.m. zones ensured that everybody (apart from landlocked states) gained something (Bailey 1997; Brown 1981; Rothwell 2012). The role of oceans in international affairs changed with the introduction and adoption of the LOS. Some hailed a new era, 'The old ocean regime ... is obsolete because the old era of ocean politics has been suspended by new patterns of conflict and alignment, and new instruments of national policy' (Osgood 1976, 10). In 1985, Booth predicted that the LOS would 'blur the boundaries between land and ocean, leading nations to feel protective and sensitive about their maritime spaces' (quoted in Baker 2013, 152). Maritime space previously dismissed as uninteresting suddenly became an entitlement in need of 'protection'.

Crucial to this expansion of the role of the ocean in international politics has been the decoupling of geophysical attributes from states' rights at sea. The 1969 North Sea cases had introduced the relevance of natural prolongation and the idea that states must take into consideration the attributes of the seabed when delineating maritime space. Then, with the conclusion of the LOS in the early 1980s, states no longer had to prove how the seabed pertained to them up until 200 n.m. to obtain rights to the resources in this domain (Kaye 2001, 20–21).

The maritime domain became legalised, internationally and under the jurisdiction of maritime states. This means a transfer of rights, from international society to the individual states, by processes unfolding on the international stage. Hurd (1999, 382) terms this the 'institutionalisation' of the norm of territorial sovereignty, which can also be applied to the oceans. '[J]ust as medieval villages were eventually fenced off in response to economic change, so states in the 1970s "fenced off" larger parts of the oceans as technological and economic change increased the uses of the oceans' (Keohane and Nye 1977, 75). Only the high seas remain a global commons.

With an international legal regime that awarded states extensive rights at sea, states themselves also became eager to uphold and defend that regime.[1] Engaging in disputes that might challenge certain aspects of specific LOS principles might prove to be a poor long-term strategy for any coastal state that benefits from these principles. Baker (2013, ii) argues that as on land, states are conditioned behaviourally by an international norm against the 'forceful acquisition of maritime spaces and resources of other states'. States have ensured a 'lock-in' of their sovereign rights at sea, while technological developments and resource demand continue to prompt greater functional use of maritime space.

Does this mean that maritime space has indeed come to take on the characteristics of territory on land? Here, it is essential to remind of the difference between land and maritime space. Occupation of the continental shelf itself could not separately lead to the acquisition of the shelf, which is contrary to sovereignty over land territory (St-Louis 2014, 16). A marked separation between land and sea also became enshrined with the LOS because rights to the latter derive from the former.

Consequently, what we are discussing regarding states and maritime space are sovereign rights to resources in the water column or on the seabed, not exclusive rights to the entire maritime 'territory' in question. States cannot deny passage through their EEZs; they may only deny actors access to marine resources and apply environmental regulations in their maritime zones. For delimitation in the maritime domain, both states may have valid legal claims to a given area, in which case it becomes a matter of 'reasonable sacrifice such as would make possible a division of the area of overlap' (Weil 1989, 91–92) or even joint sharing – as with oil and gas resources or a joint fisheries zone. We are also discussing two different forms of 'rights' by states: '... in contrast to land boundaries which separate sovereignties in their totality, maritime boundaries (with the exception of those of the territorial sea) separate only sovereign rights with a functional, and hence limited, character' (Weil 1989, 93). We must bear in mind the crucial difference between sovereign rights (EEZ, continental shelf) and complete sovereignty, as per Krasner's (1999) accounts. As Ásgeirsdóttir (2016, 190) puts it:

> Maritime boundaries differ from terrestrial boundaries in important ways. While terrestrial boundaries are often the legacy of colonial times, most current maritime boundaries have been settled by independent states. And while issues of islands and rocks in a disputed area can arouse nationalistic tendencies very few people live in the disputed areas lessening the likelihood of increased tensions. Additionally, there are often few landmarks to serve as easy reference points and there are no previously established administrative frontiers that can guide the division.

However, Weil (1989, 93–94) warns against exaggerating the difference between land and sea. Zacher (2001) argues that the value of territory (on land) can change, especially in terms of its economic significance. The same can arguably be said of maritime space. The 'entry of security considerations' into the delineating process of maritime space, as well as a general trend of 'territorialisation' of the 200 n.m. zone, Weil (1989, 93–94) notes, speaks to the growing importance of the maritime domain, as well as the expanding capacity of states to enforce and uphold their rights within this space. In recent decades, states and the international community at large have increasingly focused on ocean resources and rights. It is this growing preoccupation (by states) with maritime space that has brought the topic of maritime dispute resolution forward in international politics.

In other words, the institutionalisation of the maritime domain described here continues to prompt changes in how states engage with and perceive ocean space. From having been a 'void' to becoming a legalised space for resource exploitation and protection, the trend emanating from the maritime boundary disputes is the 'territorialisation' of the maritime domain as a continuing process.[2] This process did not come to an end with the legalisation of this domain in the 1980s: it is still underway today as states utilise more of their maritime space for resource and political purposes. In turn, maritime space and spatial rights have become the central components of the modern state.

11.2   THE OCEAN HAS SYMBOLIC QUALITIES

This process of institutionalisation (or 'territorialisation') of maritime space for states in the twenty-first century is coupled with – or fuelled by – a second trend that affects how states and their leaders view and relate to the ocean: namely, that disputes over maritime space are increasingly entangled in domestic politics. Vasquez and Valeriano (2009, 194) describe a conflict as spiralling when it becomes infused with symbolic qualities. It might be assumed that maritime disputes – whether concerned with fishing rights or boundaries – would be a simple matter of delineating rights and ownership, given the tangible character of such disputes. Huth (1998, 26), for example, has argued that 'the political salience of the [maritime] dispute is generally limited, in contrast with the importance and attention often given to land-based disputes'.

However, as shown throughout this book, when a maritime dispute reaches the political agenda, there are (domestic) actors who stand to benefit from infusing it with intangible dimensions like 'national pride' or 'being cheated out of what is ours' (Hønneland 2013).[3] Beyond institutional ratification procedures, opposition to concessions in the maritime domain takes the form of lobbying by powerful interest groups, loss of popular vote/confidence or strong media opposition (Byers and Østhagen 2017). If concessions in negotiations (inherent to any maritime boundary delimitation) are not perceived as acceptable domestically, settling the dispute will prove challenging, even if the leaders and foreign policy elites have reached an agreement through bilateral negotiations.

Whether concessions to another state are deemed acceptable will also depend on the identity of the given state: the conception of a friend or foe begins 'at home' (Hopf 2002). Maritime disputes are not devoid of the intangible and symbolic elements that can lead to conflicts escalating beyond the initial dispute itself. This concerns not only the economic interests of the actors involved, but also the wider ideas of symbolism and identity. States (and their inhabitants) do care about their maritime disputes, even those of limited economic value, and increasingly so.

For example, in bilateral Norwegian–Russian relations over fisheries around Svalbard from the 1990s onwards, the domestic audience has played a central role in these states' relations. Officials in Murmansk and representatives of the fisheries industry have attempted to infuse an intangible dimension to the underlying conflict of interests, arguing that, compared with Soviet times, Russia was now 'weak', failing to 'protect its rights' (Åtland and Ven Bruusgaard 2009; Hønneland 2013). These attempts at agenda-setting did not succeed in spurring Moscow to escalatory actions – but they show how maritime disputes are not devoid of intangible, symbolic elements that can result in conflicts escalating beyond the dispute at hand (Østhagen 2018b).

Similarly, when in 1975 Colombia agreed on its first maritime boundary with Ecuador, the government was criticised for 'wasting time' on the maritime domain and for its futile efforts at drawing an 'imaginary line in the sea' (Londoño 2015, 248).[4] Four decades later, however, in the 2018 Colombian presidential elections, the maritime boundary disputes between Nicaragua and Colombia over the San Andrés Archipelago (and the 2012 ICJ ruling) were used by the candidates to stir up popular support (Al Dia 2018). Subsequently, Nicaraguan President Ortega used the same conflict to suggest that Colombia was supporting a coup d'état, here as a way of diverting domestic efforts to get him removed from office (Rico 2018).

If the dispute in question has not attracted public attention (and domestic opposition), governments may achieve settlement, as Canada and Denmark did in 1973. However, once a dispute has become politicised, any resolution of the dispute carries domestic political risk. Indeed, even undertaking negotiations may be risky, which explains why government officials sometimes refer to negotiations as 'discussions'. These dynamics can be observed in Colombia's maritime boundary disputes with Nicaragua, Venezuela and Honduras, where public engagement in both Colombia and the opposing state effectively limited the range of options available to state leaders, as well as their ability to compromise. Any concession is seen as being directly opposed to the interests of that country. As Kleinsteiber (2013, 18) has also noted, regarding disputes in the South and East China Seas:

> While these disputes have the potential to die down if they are 'shelved' in favour of pursuing more mutually beneficial goals, they can flare up at any time, especially when driven by nationalist sentiments. This has the potential to be the troubling future of maritime conflict, when conflicts in question may be impossible to separate from national identity.

Disputes over maritime space can acquire importance in national campaigns aimed at rallying domestic support. Additionally, as states enhance their naval capabilities in line with technological developments, their capacities for monitoring and controlling their maritime zones have expanded. To a greater extent than ever before, events at sea trigger an immediate response and attention. Kleinsteiber (2013, 15) argues that '[t]he fundamental drivers behind the disputes in the East and South China Seas are not potential or claimed natural resources, but rather domestic politics, rising nationalism, and irredentism'. When in 2005 a Russian trawler 'kidnapped' two Norwegian Coast Guard officers and fled towards Russian waters after fishing in the waters around Svalbard, the Norwegian media were quick to broadcast the event live on national television, in turn helping to spur politicians into action (Fermann and Inderberg 2015). The role of maritime space in domestic politics has changed over the course of four decades – from a functional space that inspired limited engagement to that of a national space requiring 'protection' and defence.

11.3 INCREASED UTILISATION SPURS ENGAGEMENT

In conjunction with this, the function of ocean space itself has expanded, with more resources being harvested at sea, ranging from fisheries to hydrocarbons. Several trends worth highlighting are fuelling this functional expansion. The total volumes transported by sea in 2016 were 10.3 billion tonnes of cargo, more than double the 4 billion tonnes in 1990 (WTO 2019). A considerable amount of the gas that is expected to replace oil consumption will be found in offshore reservoirs (International Energy Agency (IEA) 2017), while offshore wind farms are increasingly becoming a source of global investment (Corbetta, Ho, and Pineda 2015, 7). Seabed minerals are also coming to the fore (Levin et al. 2016; Jaeckel, Gjerde, and Ardron 2017).[5] Using ship-based extraction technology, Japan successfully mined metals from its seabed in 2017 and expects large-scale commercialisation of several offshore deposits from 2020 onwards (Kyodo 2017).[6] Finally, straddling (and high seas) fish stocks constitute a shared resource,[7] but as Wood et al. (2008) emphasise, global fish stocks are decreasing because of overfishing, in international waters and within national EEZs.

On the one hand, we have the idea of the ocean and states' ocean space as a legalised, institutionalised and governed domain, where states tend to abide by the rules set forth by the LOS because it is in their common interest to do so. On the other hand, greater domestic engagement is also spurred by recognition of the ocean as a policy issue in need of common efforts to combat everything from sea-level rise to plastic pollution. We have witnessed this in the cases examined here, where maritime boundary disputes that have appeared on agendas more recently (in the past two decades) have involved a wider range of relevant factors and seem to be fostering broader public engagement than the maritime disputes tackled in the 1970s and 1980s. As put by the lead nego-tiator of Norway's latest rounds of negotiations, 'A boundary itself is just one element. More important are those normative factors increasingly related, such as military interests, economy and larger security considerations' (Norwegian Diplomat I 2017). Greater utilisation of oceans or national maritime zones in domestic politics is a trend likely to increase as ocean space continues to rise on the agenda.

When the LOS negotiations were underway, there were critical voices arguing for a 'global commons' approach to the oceans (Vogler 2000, 48–63) or that states should manage the oceans jointly (under the UN) to avoid disastrous consequences 'for the future of mankind' (Pardo 1968, 223).[8] Fuelled by the increasingly evident effects of climate change, this conception of ocean space is now also on the rise, coupled with the original/traditional view of the maritime domain as a void or domain separated from society, one faced with the effects of human activity ranging from matters directly concerned with the ocean itself to the general consequences of industrial activity, global trade and a growing global population. Greater 'territorialisation' (for exploitative purposes) will necessarily clash – on the conceptual level – with the stewardship ideas featured on the agenda today. The question is to what extent these trends are compatible. Steinberg (2001, 176) holds that we are witnessing a clash of 'social constructions' of the ocean, with the 'capitalist' or 'materialist' trends in the maritime domain outlined fuelling this clash.

The various areas of jurisdiction under the legal regime developed for the oceans are facing new challenges. Examples range from efforts to implement Marine Protected Areas (MPA) in parts of Antarctica (Østhagen and Raspotnik 2020) to the Intergovernmental Conference on an international legally binding instrument under the LOS on the conservation and sustainable use of marine biological diversity of areas beyond national jurisdiction (BBJN) (Tiller et al. 2019; Prip 2017). As shipping is increasing in territorial waters across the globe, issues of access rights, status of sea lanes and environmental protection are also at the forefront of international debates (Oude Elferink and Rothwell 2004; Jaeckel, Gjerde, and Ardron 2017). Climate change and other environmental factors are further causing variability in the spatial distribution of fish stocks, challenging established management regimes or prompting new Regional Fisheries Management Organisations (RFMOs) (Stokke 2017; 2000; Stokke, Østhagen, and Raspotnik, n.d.). In addition, the processes for determining the limits of continental shelves beyond 200 n.m. are becoming increasingly relevant and potentially conflictual (Busch 2018).

However, proponents of the LOS regime and the increasing legalisation of maritime space would argue that this offers the best framework for dealing with the issues arising over how to manage ocean space. Measures ranging from MPAs to RFMOs are indeed developed to provide mechanisms for tackling the growing number of ocean-based environmental issues. However, are these mechanisms sufficiently able to tackle the rapidly changing climatic conditions in the oceans? Are states willing to forgo potential economic benefits to deal with these challenges?

In sum, states are concerned with their maritime space and increasingly so. Also, from a purely functional perspective, maritime 'territory' has become more valuable for states. In all four countries examined in this book, maritime space has figured in some form of domestic/foreign policy agenda aimed at national development in recent decades. In all four countries, maritime disputes – settled or outstanding – have topped the policy agenda at some point in the last decade. With the sea having emerged from being literally a great blue empty space to an institutionalised policy domain, the expansion of activities taking place at sea and the growing reliance on maritime activities have resulted not only in greater importance being placed on the outcome of maritime boundary disputes, but also in shifts in the political relevance and usage of the maritime domain. Today, oceans matter more than before for states in their power relations vis-à-vis other states, as well as for political leaders seeking to sway domestic audiences.

Back in the 1960s and 1970s, when states were implementing extended maritime zones, the idea that maritime space and the location of an (invisible) and slightly arbitrary boundary at sea should hold such significance would probably have been unimaginable. However, returning to the idea of 'territorialisation', this development is not a dichotomous option of either/or: it is a process in which states' relationships to ocean space are fluctuating and shifting. The maritime domain continues to be distinct from land in terms of sovereignty and sovereign rights and, more generally, as a spatial domain, but it is also acquiring characteristics resembling those of politics over land.

It also seems reasonable to expect that as maritime space becomes increasingly relevant for states, related outstanding disputes will be more difficult to settle. The preoccupation of states and state leaders with marine resources and the general strategic value of extended maritime space, together with technological developments that enable greater control over the maritime domain (coast guard vessels, satellites, drones, subsea installations, etc.) will not render current disputes over the same space any less relevant. Changes in technology and state capacities to monitor and be present in the maritime domain may engender greater risks of conflicts emerging over ocean space. The increased use of oceans as a resource base, for everything from seabed minerals to fisheries, has further heightened the 'salience' of maritime space for states.

Additionally, as maritime disputes become infused with intangible dimensions and issues concerning symbolism and engaged domestic audiences, the characteristics of dispute containment at sea could be changing. Contrary to popular belief, maritime disputes may assume some of the same characteristics as disputes on land. Although disputes over ocean space may initially be more concerned with tangible questions of resource delimitation and 'who owns what', they can become infused with symbolism and intangible characteristics. In other words, the public interest in what happens at sea, coupled with environmental concern and economic interests under the umbrella-term the 'blue economy', leads to engagement in geopolitics at sea.

The conceptualisation of the maritime domain as a conflict reducer might be questioned. Indeed, it might be said that maritime disputes are coming to resemble traditional territorial conflicts on land. Especially in bilateral relations that are already fraught, as with Colombia–Venezuela or Norway–Russia (over Svalbard), maritime disputes may prove to hold latent conflict potential.

However, the maritime domain has certain characteristics that nevertheless keeps it separate from the terrestrial domain. There are geographical barriers that hinder prolonged interactions between the actors concerned. Maritime boundaries are also a construct of international law, and (coastal) states seem to depend on the LOS regime while also wanting to apply the regime to their own advantage.

If, as predicted by FAO (2016), fisheries continue to grow in importance in terms of livelihoods and a source of protein, certain characteristics of fisheries and maritime boundaries might also become more pronounced, spurring cooperation. As states fulfil their LOS obligation to manage transboundary fish stocks, the continued development of multinational management regimes might render the exact location of a maritime boundary less important for this purpose.

The current LOS regime was developed at a time when resource extraction from the continental shelf was gearing up. Although this matter is beyond the scope of this book, we could question to what extent this overarching regime is adequate and sufficiently adaptable to handle the changes occurring both in the oceans themselves and in the politics surrounding these processes, as environmental changes grow exponentially.

Settling maritime disputes does not seem to become an easier process with the trends described. The barriers identified here – like domestic audiences and security relations – will remain relevant, perhaps even more so. Moreover, potential distrust or the limited utilisation of international adjudication and arbitration might not make it easier to settle maritime boundaries. On the other hand, the use of complex resource-sharing mechanisms or the increasing focus on developing adequate RFMOs concerned with various transboundary fish stocks, as well as the establishment of MPAs in tandem with greater environmental awareness concerning the state of the oceans, might make the exact location of the maritime boundary itself (if not the maritime domain) less important.

Establishing agreements on these mechanisms is still necessary but perhaps with a slightly different focus than when settling maritime boundaries in the traditional sense. Managing the disputed maritime area might also be an easier, even preferred, solution in some instances. That being said, it does not seem likely that maritime boundaries – settled or not – and related issues of resource management, ownership and access are likely to become less relevant in the years to come.

11.4   CONCLUDING REMARKS

When the international community agreed on an international legal framework for the oceans, all coastal states were granted the right to extended maritime zones. This innovation did not only secure sovereign rights to the resources on the seabed and in the water column further offshore than thought possible a few decades earlier, but it also created a problem. Suddenly, the oceans were being carved up into various national zones, just as European states had carved up 'free' land on other continents centuries earlier. In contrast to delineating borders on land, this new ocean imperialism was based on geometric propositions, the shape of the relevant coastline and international law.

Although it served as the impetus for implementing extended maritime zones across the world, the LOS Convention did nothing to help solve the problem that arose as a consequence: overlapping maritime claims and boundary disputes between states. In some instances, states managed to agree rather quickly on where to delineate the maritime boundary between opposing or adjacent coastlines, with the issue coming to fore on national agendas in the period from 1960 to 1980. However, most boundaries at sea were not agreed in that period. Some took years of negotiations to settle. Some lay dormant for decades before suddenly emerging in the limelight. Many still remain in dispute.

Although always of existential importance to humans, the oceans have been on a slow political ascent over the past few decades. Environmental concern and engagement, coupled with greater utilisation of the oceans for economic purposes through resource exploitation and transport, have propelled previously dormant or inconsequential maritime disputes onto the political agenda. At the outset, it seems that the benefits of agreeing on a clear boundary marking the jurisdiction and sovereign rights of each coastal state would greatly outweigh the costs of the concessions made through negotiations. However, this has not been the case, and almost half of all maritime boundaries remain disputed. This book has examined why states settle their maritime boundary disputes and why some of them do not, along with what this means for geopolitics at sea.

First, it is not sufficient to study maritime boundary disputes solely as individual cases. How each dispute relates to other disputes involving the same countries and in the same maritime region must be taken into consideration. Second, we need to conceive of the outcome beyond a binary settled/ not settled. By depicting the outcome as a process with various steps, we can understand the nuances in what drives and hinders the settlement of maritime boundary disputes at different points in that process.

Moreover, this book has unpacked two apparent contradictions when it comes to why states settle maritime boundary disputes. It has been argued that the presence of oil and gas resources can enable or hinder states when settling disputes at sea. Fisheries have similarly been portrayed as a potential barrier to boundary agreements, or it has been discarded as having limited relevance. A more nuanced and complex picture has emerged here by exploring these causal logics across cases, where I have found other variables that condition the effect of hydrocarbon and living marine resources on boundary disputes. Not even the allure of oil and gas profits is sufficient to overcome legal and domestic barriers when these are present or to obliterate the legacy of negative historic relations with the opposing state.

What is further clear is the importance of the legal characteristics of a maritime boundary dispute. This is hardly surprising because a boundary at sea is a legal construct in and of itself (in contrast to, say, borders between states due to the separating force of a river or a mountain chain). What is novel is the attempt to view legal characteristics in conjunction with other political explanatory factors, in essence using international law to help explain a complex outcome in IR. A maritime boundary is inherently a compromise whereby each state usually concedes at least some part of its maximum zonal claim. 'Hard points' such as treaties or court/arbitration decisions before the arrival of the LOS confer less flexibility when negotiating positions (Byers and Østhagen 2017). The same goes for a state's concern over legal precedent when the dispute at hand is seen in conjunction with other outstanding disputes. Concern for precedent can also – as seen in some cases here – serve as an enabler for settling disputes.

Finally, the domestic dimension is a neglected aspect of ocean boundary settlement processes, one that this book has explored. Separating this aspect into three different strands – ratification procedures, public engagement and economic interests – has shown that the domestic audience component of maritime disputes is not only highly relevant to understanding why settlement efforts are halted or not ratified: it is also becoming increasingly important.

States and their political leaders at times deliberately focus on the ocean and related disputes. This means that states are not bound to look towards the sea as a response to systemic changes; they choose to do so to advance their own interests. Canada put ocean issues at the top of the agenda during its Group of Seven (G7) Presidency in 2018. Norway has chosen to focus on ocean growth from 2017 onwards as a way of transcending its dependency on oil and gas exports. In other instances, previously 'managed' or 'frozen' boundary disputes demand attention. Australia's dispute with Timor-Leste was forced on the agenda by public discontent in 2017. The maritime dispute with Nicaragua was centre stage in Colombia's 2018 presidential election. In the summer of 2019 and 2020, the drilling for oil and gas by Turkish vessels in Cypriot waters propelled the larger dispute over Cyprus, as well as the multiple unresolved boundary disputes in the eastern Mediterranean, onto the international agenda.

This speaks to the growing importance of maritime space for states and their interest groups, as well as the infusion of symbolic and/or intangible elements into boundary negotiations – neither of which was particularly prominent in the 1960s and 1970s. Moreover, the domestic audience component of an initially legal and technical exercise might make it difficult for states to yield maritime space in bilateral negotiations. Thus, a trend emerging throughout this book is the 'politicisation' of the ocean – or the 'territorialisation' of maritime space.

The expansion in human utilisation of the oceans because of transportation and resource needs, as well as a greater awareness of what is occurring at sea from both an economic and science perspective, has resulted in greater attention and value being given to maritime space. Maritime boundaries and related disputes are likely to continue to appear on the political agenda. This also brings with it the realisation that the specific location of the boundary might sometimes be less important when the mechanisms for resource sharing can be used.

These points are relevant not only for straddling hydrocarbons or fish stocks, but also when managing the latent conflict potential between states at sea. As states and companies continue to develop technology to further expand their ability to utilise ocean space and resources, the question of maritime rights and responsibilities will only become more important to human society. As the author Arthur C. Clarke succinctly put it, 'How inappropriate to call this planet Earth, when clearly it is Ocean'.[9]

NOTES

1. See, for example, the studies of Arctic maritime boundaries and disputes (Moe, Fjærtoft, and Øverland 2011; Tamnes and Offerdal 2014; Byers 2017; Claes and Moe 2018).
2. See, for example, Elden 2013; Agnew 1994; Storey 2012. Territoriality can be defined as the process whereby territory (here: the ocean) is claimed by individuals or groups.
3. Interestingly, this was a prominent part of the campaign to leave the EU during the BREXIT referendum in the UK in 2016, despite fisheries only accounting for 0.05% of the country's GDP (Lichfield 2018).
4. Author's translation, with assistance from Victoria Dalgleish Lindbak.
5. By 2017, the International Seabed Authority had granted 28 contracts to mining companies, giving them exclusive rights to explore large parts of the continental shelf *beyond* national jurisdiction in the Pacific, Atlantic and Indian Oceans (Woody 2017). However, there are serious concerns about the environmental impacts of such activities and the lack of transparent and sufficient regulation (Jaeckel 2016; Levin et al. 2016), along with calls to develop an 'exploitation code regulation' by 2020 (Woody 2017).
6. Other potentially valuable deposits have recently been discovered across the world ocean. In the Atlantic, British scientists discovered minerals that 'contain

the scarce substance tellurium in concentrations 50,000 times higher than in deposits on land' (Shukman 2017).
7. As an object that cannot be appropriated to any individual group and, therefore, is vulnerable to overexploitation by self-interested/uncoordinated individual profit-seeking behaviour (the classic 'tragedy of the commons') (Crowe 1969, 1103–4; Stuart 2013).
8. However, argumentation was used during the LOS negotiations in the 1970s that favoured the economic interests of states while using language that emphasised environmental concerns: 'Within the framework of the [LOS] Conference, Latin Americans, Africans, Arabs, and Asians found in some Western nations, led by environment conscious Canada and Norway, strong allies that began to rationalize and sell the EZ concept in language well understood by the maritime powers of the West and at least not misunderstood by the Soviet Union' (Nweihed, 1980, p. 7).
9. James Lovelock quoting Arthur C. Clark, in his book *Gaia: A New Look at Life on Earth* from 1979 (Oxford University Press), p. 78.

Bibliography

Adler, Emanuel, and Michael Barnett. 1998. *Security Communities*. Edited by Emanuel Adler and Michael Barnett. Cambridge: Cambridge University Press.

Agnew, John. 1994. "The Territorial Trap: The Geographical Assumptions of International Relations Theory." *Review of International Political Economy* 1 (1): 53–80. https://doi.org/10.1080/09692299408434268.

Akami, Tomoko, and Anthony Milner. 2013. "Australia in the Asia-Pacific Region." In *The Cambridge History of Australia: Volume 2 – The Commonwealth of Australia*, edited by Alison Bashford and Stuart Macintyre, 537–60. Port Melbourne: Cambridge University Press.

Al Dia. 2018. "Ties with U.S., Nicaragua Dominate First Colombia Presidential Debate." *Al Dia*, April 20, 2018. http://aldianews.com/articles/politics/elections/ties-us-nicaragua-dominate-first-colombia-presidential-debate/52393.

Alexander, Lewis M. 1993. "Canada–Denmark (Greenland)." In *International Maritime Boundaries Vol. 1–2*, edited by Jonathan I. Charney and Lewis M. Alexander, 371–78. Dordrecht: Martinus Nijhoff.

Allee, Todd, and Paul Huth. 2006. "The Pursuit of Legal Settlements to Territorial Disputes." *Conflict Management and Peace Science* 23 (4): 285–307.

Allison, Edward H., Allison L. Perry, Marie Caroline Badjeck, W. Neil Adger, Katrina Brown, Declan Conway, Ashley S. Halls et al. 2009. "Vulnerability of National Economies to the Impacts of Climate Change on Fisheries." *Fish and Fisheries* 10 (2): 173–96.

Anand, Ram Prakash. 1983. *Origin and Development of the Law of the Sea: History of International Law Revisited*. The Hague: Martinus Nijhoff.

Anderson, D. H. 2009. "The Status Under International Law of the Maritime Areas Around Svalbard." *Ocean Development & International Law* 40 (4): 373–84.

Applebaum, B. 2001. "Law of the Sea – The Exclusive Economic Zone An Overview of the Development of the 200 Mile Exclusive Economic Zone and the Related Canadian Fisheries and Other Interests." Canadian Museum of History. https://www.historymuseum.ca/cmc/exhibitions/hist/lifelines/apple1e.shtml#contents.

Aréchaga, Eduardo Jiménez de. 1993. "Colombia–Ecuador." In *International Maritime Boundaries Vol. 1*, edited by Jonathan I. Charney and Lewis M. Alexander, 809–13. Dordrecht: Martinus Nijhoff.

Árnadóttir, Snjólaug. 2016. "Termination of Maritime Boundaries Due to a Fundamental Change of Circumstances." *Utrecht Journal of International and European Law* 32: 94–111.

Ásgeirsdóttir, Áslaug. 2016. "Settling of the Maritime Boundaries of the United States: Cost of Settlement and the Benefits of Legal Certainty." *Marine Policy* 73: 187–95.

Ásgeirsdóttir, Áslaug, and Martin C. Steinwand. 2015. "Dispute Settlement Mechanisms and Maritime Boundary Settlements." *Review of International Organizations* 10: 119–43.

Ásgeirsdóttir, Áslaug, and Martin C. Steinwand. 2016. "Distributive Outcomes in Contested Maritime Areas." *Journal of Conflict Resolution* 62 (6): 1284–1313.

Åtland, Kristian, and Kristin Ven Bruusgaard. 2009. "When Security Speech Acts Misfire: Russia and the Elektron Incident." *Security Dialogue* 40 (3): 333–53.

Australia–East Timor. 2002. *Timor Sea Treaty*. http://timor-leste.gov.tl/wp-content/uploads/2010/03/R_2003_2-Timor-Treaty.pdf.

Australia–France. 1982. *Agreement on Maritime Delimitation (with Maps). Signed at Melbourne on 4 January 1982*. http://www.marineregions.org/documents/volume-1329-I-22302-English.pdf.

Australia–Indonesia. 1971. *Agreement between the Government of the Commonwealth of Australia and the Government of the Republic of Indonesia Establishing Certain Seabed Boundaries (18 May 1971)*.

Australia–Indonesia. 1973. *Agreements between Australia and Indonesia Concerning Certain Boundaries between Papua New Guinea and Indonesia*. http://www.un.org/Depts/los/LEGISLATIONANDTREATIES/PDFFILES/TREATIES/AUS-IDN1973PNG.pdf.

Australia–Indonesia. 1974. *Memorandum of Understanding between the Government of Australia and the Government of the Republic of Indonesia Regarding the Operations of Indonesian Traditional Fishermen in Areas of the Australian Exclusive Fishing Zone and Continental Shelf, 1974*.

Australia–Indonesia. 1989. *Treaty between Australia and the Republic of Indonesia on the Zone of Cooperation in an Area between the Indonesian Province of East Timor and Northern Australia*. http://www.austlii.edu.au/au/other/dfat/treaties/1991/9.html.

Australia–Indonesia. 1997. *Treaty between the Government of Australia and the Government of the Republic of Indonesia Establishing an Exclusive Economic Zone Boundary and Certain Seabed Boundaries*. http://www.un.org/Depts/los/LEGISLATIONANDTREATIES/PDFFILES/TREATIES/AUS-IDN1997EEZ.pdf.

Australia–Papua New Guinea. 1978. *Treaty between Australia and the Independent State of Papua New Guinea Concerning Sovereignty and Maritime Boundaries in the Area between the Two Countries, Including the Area Known as Torres Strait, And Related Matters*. http://www.un.org/Depts/los/LEGISLATIONANDTREATIES/PDFFILES/TREATIES/AUS-PNG1978TS.PDF.

Australia–Solomon Islands. 1988. *Agreement between the Government of Solomon Islands and the Government of Australia Establishing Certain Sea and Seabed Boundaries*. http://www.un.org/Depts/los/LEGISLATIONANDTREATIES/PDFFILES/TREATIES/aus-sol1988.tif.

Australia–Timor-Leste. 2006. *Treaty between Australia and the Democratic Republic of Timor-Leste on Certain Maritime Arrangements in the Timor Sea (CMATS)*. http://www.austlii.edu.au/au/other/dfat/treaties/2007/12.html.

Australia–Timor-Leste. 2018. *Treaty Between Australia and the Democratic Republic of Timor-Leste Establishing Their Maritime Boundaries in the Timor Sea*. https://dfat.gov.au/geo/timor-leste/Documents/treaty-maritime-arrangements-australia-timor-leste.pdf.

Australian Government Official. 2019. "Foreign Ministry of Australia, Australian Government" interviewed by Andreas Østhagen, Canberra, February 2019.

Bailey, James E. 1997. "The Unanticipated Effects of Boundaries: The Exclusive Economic Zone and Geographically Disadvantaged States Under UNCLOS III." *Boundary and Security Bulletin* 5 (1): 87–95. http://www.dur.ac.uk/ibru/publications/view/?id=106.

Baker, James S. 2013. *International Order in the Oceans: Territoriality, Security and the Political Construction of Jurisdiction over Resources at Sea*. Vancouver: University of British Columbia. http://legacy.politics.ubc.ca/11946/.

Baker, James S., and Michael Byers. 2012. "Crossed Lines: The Curious Case of the Beaufort Sea Maritime Boundary Dispute." *Ocean Development & International Law* 43 (March 2010): 70–95.

Bakken, Laila Ø., and Kristian Aanensen. 2010. "- Historisk løsning av delelinjen." NRK, April 27, 2010.

Bateman, Sam. 2007. "UNCLOS and its Limitations as the Foundation for a Regional Maritime Security Regime." *Korean Journal of Defense Analysis* 19 (3): 27–56.

BBC News. 2012. "Colombia Pulls out of International Court over Nicaragua." *BBC News*, November 28, 2012. https://www.bbc.com/news/world-latin-america -20533659.

Beckman, Robert, and Clive Schofield. 2009. "Moving Beyond Disputes Over Island Sovereignty: ICJ Decision Sets Stage for Maritime Boundary Delimitation in the Singapore Strait." *Ocean Development & International Law* 40 (1): 1–35. https://doi .org/10.1080/00908320802631551.

Beckman, Robert, Ian Townsend-Gault, Clive Schofield, Tara Davenport, and Leonardo Bernard, eds. 2013. *Beyond Territorial Disputes in the South China Sea: Legal Frameworks for the Joint Development of Hydrocarbon Resources*. Cheltenham, UK and Northampton, MA, USA: Edward Elgar Publishing.

Benton, Lauren. 2010. *A Search for Sovereignty: Law and Geography in European Empires, 1400–1900*. New York: Cambridge University Press.

Berakit, Tanjung. 2011. "Mending the Imaginary Wall between Indonesia and Malaysia." *Wacana* 13 (1): 1–28.

Bernhofen, Daniel M., Zouheir El-Sahli, and Richard Kneller. 2016. "Estimating the Effects of the Container Revolution on World Trade." *Journal of International Economics* 98: 36–50. https://doi.org/10.1016/j.jinteco.2015.09.001.

Bissinger, Jared. 2010. "The Maritime Boundary Dispute between Bangladesh and Myanmar: Motivations, Potential Solutions, and Implications." *Asia Policy* 10 (1): 103–42.

Blake, Gerald H. 1994. *Maritime Boundaries: World Boundaries Volume 5*. London: Routledge.

Blomqvist, Lene Buer. 2006. "Har Norge Behov for En Avklart Delelinje i Barentshavet? – En Diskursanalyse Av Delelinjekonflikten Mellom Norge Og Russland (Does Norway Have a Need to Settle the Boundary in the Barents Sea? – A Discourse Analysis of the Boundary Dispute between Norway)." University of Tromsø, Norway.

Booth, Ken. 1985. *Law, Force and Diplomacy at Sea*. Abingdon: Routledge.

Boswell, Randy. 2008. "Astonishing' Data Boost Arctic Claim." *Ottawa Citizen*, November 12, 2008.

Bourne, Charles B., and Donald M. McRae. 1976. "Maritime Jurisdiction in the Dixon Entrance: The Alaska Boundary Re-Examined." *Canadian Yearbook of International Law* 14: 175–223.

Brabandere, Eric De. 2016. "The Use of Precedent and External Case-Law by the International Court of Justice and the International Tribunal for the Law of the Sea." *Grotius Centre Working Paper* 57: 24–55.

Brander, Keith. 2010. "Impacts of Climate Change on Fisheries." *Journal of Marine Systems* 79 (3–4): 389–402.

British Columbia. 1977. *Submission of the Province of British Columbia on West Coast Maritime Boundaries between Canada and the United States*. Victoria: Queen's Printer.

Brooks, Stephen. 2010. "Canada–United States Relations." In *The Oxford Handbook of Canadian Politics*, edited by John C. Courtney and David E. Smith, 378–394. Oxford: Oxford University Press.

Brown, E. D. 1981. "Delimitation of Offshore Areas. Hard Labour and Bitter Fruits at UNCLOS III." *Marine Policy* 5 (3): 172–84.

Brownlie, Ian. 1980. *Legal Status of Natural Resources in International Law (Some Aspects). Volume 162 of Académie de Droit International de La Haye. Recueil Des Cours*. Alphen aan der Rijn: Sijthoff et Noordhoff.

Brunnée, Jutta, and Stephen J. Toope. 2010. *Legitimacy and Legality in International Law*. Cambridge: Cambridge University Press.

Buchanan, Allen, and Margaret Moore, eds. 2003. *States, Nations and Borders: The Ethics of Making Boundaries*. Cambridge: Cambridge University Press.

Bull, Hedley. 1976. "Sea Power and Political Influence." *The Adelphi Papers* 16 (122): 1–9.

Burmester, Henry. 1982. "The Torres Strait Treaty: Ocean Boundary Delimitation by Agreement." *The American Journal of International Law* 76 (2): 321–49.

Burmester, Henry. 1995. "Australia and the Law of the Sea." In *The Law of the Sea in the Asian Pacific Region*, edited by James Crawford and Donald R. Rothwell, 51–64. Leiden: Kluwer Academic.

Burmester, Henry. 2019. "Former Lead Negotiator, Attorney-General's Department, Australian Government" interviewed by Andreas Østhagen, Canberra, January 2019.

Busch, Signe Veierud. 2018. "The Delimitation of the Continental Shelf beyond 200 Nm: Procedural Issues." In *Maritime Boundary Delimitation: The Case Law – Is It Consistent and Predictable?*, edited by Alex G. Oude Elferink, Tore Henriksen, and Signe Veierud Busch, 319–50. Cambridge: Cambridge University Press.

Buzan, Barry, and Ole Wæver. 2003. *Regions and Powers: The Structure of International Security*. Cambridge: Cambridge University Press.

Byers, Michael. 1999a. *Custom, Power and the Power of Rules: International Relations and Customary International Law*. Cambridge: Cambridge University Press.

Byers, Michael. 1999b. *Custom, Power and the Power of Rules*. Cambridge: Cambridge University Press.

Byers, Michael. ed. 2000. *The Role of Law in International Politics: Essays in International Relations and International Law*. Oxford: Oxford University Press.

Byers, Michael. 2009. *Who Owns the Arctic*. Vancouver: Douglas & McIntyre.

Byers, Michael. 2013. *International Law and the Arctic*. New York: Cambridge University Press.

Byers, Michael. 2017. "Crises and International Cooperation: An Arctic Case Study." *International Relations* 31 (4): 375–402.

Byers, Michael, and Suzanne Lalonde. 2009. "Who Controls the Northwest Passage?" *Vanderbilt Journal of Transnational Law*. https://doi.org/10.1080/02722018809480935.

Byers, Michael, and Andreas Østhagen. 2017. "Why Does Canada Have So Many Unresolved Maritime Boundary Disputes?" *Canadian Yearbook of International Law* 54 (October): 1–62.

Byers, Michael, and Andreas Østhagen. 2018. "Settling Maritime Boundaries: Why Some Countries Find It Easy, and Others Do Not." In *The Future of Ocean*

Governance and Capacity Development, edited by The International Ocean Institute-Canada, 162–68. Leiden: Brill Nijhoff.

Byers, Michael, and Stewart Webb. 2013. "Titanic Blunder: Arctic/Offshore Patrol Ships on Course for Disaster." Ottawa.

Calderbank, Bruce, Alec M. MacLeod, David H. Gray, and Ted L. McDorman. 2006. *Canada's Offshore: Jurisdiction, Rights, and Management.* Victoria: Trafford Publishing.

Campbell, David. 1998. *Writing Security: United States Foreign Policy and the Politics of Identity.* Minneapolis: University of Minnesota Press.

Canada–Denmark. 1973. *Agreement Relating to the Delimitation of the Continental Shelf between Greenland and Canada (with Annexes). Signed at Ottawa on 17 December 1973.* https://treaties.un.org/doc/Publication/UNTS/Volume 950/ volume-950-I-13550-English.pdf.

Canada–France. 1972. *Agreement between the Government of Canada and the Government of France on Their Mutual Fishing Relations.*

Canada–France. 1992. *Court of Arbitration for the Delimitation of Maritime Areas between Canada and France: Decision in Case Concerning Delimitation of Maritime Areas (St. Pierre and Miquelon).*

Canada–United States. 1846. *Treaty Establishing the Boundary in the Territory on the Northwest Coast of America Lying Westward of the Rocky Mountains.* https:// web.archive.org/web/20091113034143/http://www.lexum.umontreal.ca/ca_us/en/ cus.1846.28.en.html.

Canada–United States. 1903. *Convention between His Majesty and the United States of America, for the Adjustment of the Boundary between the Dominium of Canada and the Territory of Alaska.*

Canadian Department of Foreign Affairs. 2012. "Canada and Kingdom of Denmark Reach Tentative Agreement on Lincoln Sea Boundary." http://news.gc.ca/web/ article-en.do?nid=709479.

Caro, Carlo Jose Vicente. 2017. "Nicolas Maduro Wants War With Colombia." *Forbes*, April 11, 2017.

Caron, David D. 2009. "Climate Change, Sea Level Rise and the Coming Uncertainty in Oceanic Boundaries: A Proposal to Avoid Conflict." In *Maritime Boundary Disputes, Settlement Processes, and the Law of the Sea*, edited by Seoung-Yong Hong and Jon M. Van Dyke, 2–17. Leiden: Martinus Nijhoff.

Carter, David B. 2010. "The Strategy of Territorial Conflict." *American Journal of Political Science* 54 (4): 969–87.

Casey, Nicholas, and Jenny Carolina González. 2019. "A Staggering Exodus: Millions of Venezuelans Are Leaving the Country, on Foot." *The New York Times*, February 20, 2019.

Charney, Jonathan I., and Robert W. Smith. 2002. "Colombia–Costa Rica." In *International Maritime Boundaries Vol. 4*, edited by Jonathan I. Charney and Robert W. Smith, 2641. Dordrecht: Martinus Nijhoff.

Checkel, Jeffrey T. 1998. "Review: The Constructivist Turn in International Relations Theory." *World Politics* 50 (2): 324–48.

Checkel, Jeffrey T. 2008. "Constructivism and Foreign Policy." In *Foreign Policy: Theories, Actors, Cases*, edited by Steve Smith, Amelia Hadfield, and Tim Dunne, 71–80. Oxford: Oxford University Press.

Cheng, Joseph Y.S., and Stephanie Paladini. 2014. "China's Ocean Development Strategy and Its Handling of the Territorial Conflicts in the South China Sea."

Philippine Political Science Journal 35 (2): 185–202. https://doi.org/10.1080/01154451.2014.965476.

Cheung, William W. L., Miranda C. Jones, Vicky W. Y. Lam, Dana D. Miller, Yoshitaka Ota, Louise Teh, and Ussif R. Sumaila. 2017. "Transform High Seas Management to Build Climate Resilience in Marine Seafood Supply." *Fish and Fisheries* 18 (2): 254–63.

Chile, Ecuador, and Peru. 1952. *Declaration on the Maritime Zone.* Chile: United Nations. https://treaties.un.org/doc/Publication/UNTS/volume 1006/volume-1006-I-14758-English.pdf.

Churchill, Robin R. 1994. "The Greenland–Jan Mayen Case and its Significance for the International Law of Maritime Boundary Delimitation." *International Journal of Marine and Coastal Law* 9 (1): 1.

Churchill, Robin R. 2001. "Claims to Maritime Zones in the Arctic – Law of the Sea Normality or Polar Peculiarity?" In *The Law of the Sea and Polar Maritime Delimitation and Jurisdiction*, edited by Alex G Oude Elferink and Donald R Rothwell, 105–24. The Hague: Martinus Nijhoff.

Churchill, Robin, and Geir Ulfstein. 1992. *Marine Management in Disputed Areas: The Case of the Barents Sea.* London: Routledge.

Churchill, Robin, and Geir Ulfstein. 2010. "The Disputed Maritime Zones around Svalbard." In *Changes in the Arctic Environment and the Law of the Sea*, edited by Myron H. Nordquist, John Norton Moore, and Thomas H. Heidar, 551–93. Leiden: Martinus Nijhoff.

Claes, Dag H., and Arild Moe. 2018. "Arctic Offshore Petroleum: Resources and Political Fundamentals." In *Arctic Governance: Energy, Living Marine Resources and Shipping*, edited by Svein Vigeland Rottem, Ida Folkestad Soltvedt, and Geir Hønneland, 9–26f. London: I. B. Tauris.

Cobb, Roger W., and Charles D. Elder. 1972. *Participation in American Politics: The Dynamics of Agenda-Building.* Boston: Allyn & Bacon.

Cobb, Roger W., J. K. Ross, and M. H. Ross. 1976. "Agenda Building as a Comparative Political Process." *American Political Science Review* 70 (1): 126–38.

Cohen, Harlan Grant. 2015. "Theorizing Precedent in International Law." In *Interpretation in International Law*, edited by Andrea Bianchi, Daniel Peat, and Matthew Windsor, 268–89. Oxford: Oxford University Press.

Colombia–Costa Rica. 1977. *Treaty on Delimitation of Marine and Submarine Areas and Maritime Co-Operation between the Republic of Colombia and the Republic of Costa Rica.*

Colombia–Dominican Republic. 1978. *Agreement on Delimitation of Marine and Submarine Areas and Maritime Cooperation between the Republic of Colombia and the Dominican Republic.*

Colombia–Ecuador. 1975. *Agreement Concerning Delimitation of Marine and Submarine Areas and Maritime Cooperation between the Republics of Colombia and Ecuador.*

Colombia–Haiti. 1978. *Agreement on the Delimitation of the Maritime Boundaries between Colombia and Haiti.*

Colombia–Honduras. 1986. *Maritime Delimitation Treaty between Colombia and Honduras.*

Colombia–Jamaica. 1993. *Maritime Delimitation Treaty between Jamaica and the Republic of Colombia.*

Colombia–Panama. 1976. *Treaty on the Delimitation of Marine and Submarine Areas and Related Matters between the Republic of Panama and the Republic of Colombia.*

Colombia–Venezuela. 1941. *Border Demarcation Agreement and Navigation of the Common Rivers between Colombia and Venezuela.*

Conley, Heather A., Matthew Melino, and Andreas Østhagen. 2017. *Maritime Futures: The Arctic and the Bering Strait Region.* Lanham: Rowman & Littlefield.

Cook, Beverly. 2005. "Lobster Boat Diplomacy: The Canada–US Grey Zone." *Marine Policy* 29 (5): 385–90. https://doi.org/10.1016/j.marpol.2004.05.010.

Corbetta, Giorgio, Andrew Ho, and Iván Pineda. 2015. "Wind Energy Scenarios for 2030." https://www.ewea.org/fileadmin/files/library/publications/reports/EWEA-Wind-energy-scenarios-2030.pdf.

Cottier, Thomas. 2015. *Equitable Principles of Maritime Boundary Delimitation: The Quest for Distributive Justice in International Law.* Cambridge: Cambridge University Press.

Crowe, Beryl L. 1969. "The Tragedy of the Commons Revisited." *Science* 166 (3909): 1103–7.

Cui, Shunji. 2014. "Conflict Transformation: The East China Sea Dispute and Lessons from the Ecuador–Peru Border Dispute." *Asian Perspective* 38 (2): 285–310.

Dagens Næringsliv. 2005. "Norsk-russisk sameie av sone (Norwegian-Russian Co-Ownership of Zone)." Dagens Næringsliv, 2005.

Davenport, Tara. 2013. "Southeast Asian Approaches to Maritime Boundaries." *Asian Journal of International Law* 4 (2): 309–57. https://doi.org/10.1017/S2044251313000313.

Denmark–Norway. 2006. *Agreement between the Government of the Kingdom of Norway on the One Hand, and the Government of the Kingdom of Denmark Together with the Home Rule Government of Greenland on the Other Hand, Concerning the Delimitation of the Continental Shelf and the Fis.*

Devabhaktuni, Sai S., and Gregory Kennedy. 2012. "Global Shipping: Any Port in a Storm? (PIMCO)." *PIMCO.* http://advisoranalyst.com/glablog/2012/05/21/global-shipping-any-port-in-a-storm.html/.

Diehl, Paul F., Derrick V. Frazier, Todd L. Allee, and Shannon O'Lear. 2006. "Introduction to CMPS Special Issue on Territorial Conflict Management." *Conflict Management and Peace Science* 23 (4): 263–65. https://doi.org/10.1080/07388940600972610.

Doherty, Ben. 2017. "Australia and Timor-Leste to Negotiate Permanent Maritime Boundary." *The Guardian*, January 9, 2017.

Downs, Anthony. 1972. "Up and Down with Ecology: The 'Issue-Attention Cycle.'" *The Public Interest* 28: 38–50.

Dunn, James. 2003. *East Timor: A Rough Passage to Independence.* Haberfield: Longueville Books.

Economist. 2006. "Treasure on the Ocean Floor." *Economist*, December 2, 2006. http://www.economist.com/node/8312172.

Economist. 2010. "Ocean Diversity: What Lies Beneath." *Economist*, August 5, 2010. https://www.economist.com/news/2010/08/05/what-lies-beneath.

Economist. 2011. "Seasteading: Cities on the Ocean." *Economist*, December 3, 2011. https://www.economist.com/printedition/specialreports/prb/sr/techstartups?year%5Bvalue%5D%5Byear%5D=2011&category=All.

Economist. 2012. "Colombia and Nicaragua: Hot Waters." *Economist*, November 29, 2012. https://www.economist.com/blogs/americasview/2012/11/colombia-and-nicaragua.

Economist. 2013. "The Future of the Oceans: Acid Test." *Economist*, November 21, 2013. https://www.economist.com/science-and-technology/2013/11/21/acid-test.

Economist. 2014a. "Ocean Acidification: A Shrinking Problem." *Economist*, January 18, 2014. https://www.economist.com/science-and-technology/2014/01/18/a-shrinking-problem.

Economist. 2014b. "Governing the Oceans: The Tragedy of the High Seas." *Economist*, February 22, 2014. https://www.economist.com/leaders/2014/02/22/the-tragedy-of-the-high-seas.

Economist. 2016a. "Marine Conservation: Rejuvenating Reefs." *Economist*, 2016. https://www.economist.com/international/2016/02/13/rejuvenating-reefs.

Economist. 2016b. "Banyan: Trawling for Trouble." *Economist*, April 14, 2016.

Economist. 2017a. "Plucking Minerals from the Seabed is Back on the Agenda." *Economist*, February 2017. https://www.economist.com/news/science-and-technology/21717351-fruits-de-mer-plucking-minerals-seabed-back-agenda.

Economist. 2017b. "The Mesopelagic: Cinderella of the Oceans." *Economist*, April 15, 2017. https://www.economist.com/science-and-technology/2017/04/15/the-mesopelagic-cinderella-of-the-oceans.

Economist. 2017c. "Deep Trouble: How to Improve the Health of the Ocean." *Economist*, May 27, 2017. https://www.economist.com/leaders/2017/05/27/how-to-improve-the-health-of-the-ocean.

Economist. 2017d. "Getting Serious About Overfishing." *Economist*, May 27, 2017. http://www.economist.com/news/briefing/21722629-oceans-face-dire-threats-better-regulated-fisheries-would-help-getting-serious-about.

Economist. 2017e. "Improving the Ocean: Getting Serious about Overfishing." *Economist*, May 27, 2017. https://www.economist.com/briefing/2017/05/26/getting-serious-about-overfishing.

Eiran, Ehud. 2017. "Between Land and Sea: Spaces and Conflict Intensity." *Territory, Politics, Governance* 5 (2): 190–206. https://doi.org/10.1080/21622671.2016.1265466.

Elden, Stuart. 2013. *The Birth of Territory*. Chicago: University of Chicago Press.

Emmers, R. 2010. "The Changing Power Distribution in the South China Sea: Implications for Conflict Management and Avoidance." *Political Science* 62 (2): 118–31.

FAO. 2014. *The State of World Fisheries and Aquaculture. Food and Agriculture Oraganization of the United Nations*. Rome: Food and Agriculture Organization of the United Nations. http://scholar.google.com/scholar?hl=en&btnG=Search&q=intitle:THE+STATE+OF+WORLD+FISHERIES+AND+AQUACULTURE#0.

FAO. 2016. *The State of the World Fisheries and Aquaculture: Contributing to Food Security and Nutrition for All*. Rome: Food and Agriculture Organization of the United Nations. http://www.fao.org/3/a-i5555e.pdf.

FAO. 2018. *The State of World Fisheries and Aquaculture*. Rome. http://www.fao.org/3/CA0190EN/ca0190en.pdf.

Fearon, James D. 1995. "Rationalist Explanations for War." *International Organization* 49 (3): 379–414.

Fermann, Gunnar, and Tor Håkon Inderberg. 2015. "Norway and the 2005 Elektron Affair: Conflict of Competencies and Competent Realpolitik." In *War: An Introduction to Theories and Research on Collective Violence*, edited by Tor Georg Jakobsen, 2nd ed., 373–402. New York: Nova Science.

Fietta, Stephen, and Robin Cleverly. 2016. *A Practitioner's Guide to Maritime Boundary Delimitation*. Oxford: Oxford University Press.

Filippone, Renee. 2015. "Shell Ends Exploration in Arctic Near Alaska 'for the Foreseeable Future.'" *CBC News*, September 28, 2015. http://www.cbc.ca/news/business/shell-stops-arctic-drilling-development-1.3246355.

Finnemore, Martha, and Kathryn Sikkink. 1998. "International Norm Dynamics and Political Change." *International Organization* 52 (4): 887–917.

Finnemore, Martha, and Stephen J. Toope. 2001. "Alternatives to 'Legalization': Richer Views of Law and Politics." *International Organization* 55 (3): 743–58.

Fjærtoft, Daniel, Moe Arild, Natalia Smirnova, and Alexey Cherepovitsyn. 2018. "Unitization of Petroleum Fields in the Barents Sea: Towards a Common Understanding?" *Arctic Review on Law and Politics* 9: 72–96. https://doi.org/10.23865/arctic.v9.1083.

Forbes, Vivian Louis. 1995. *The Maritime Boundaries of the Indian Ocean Region*. Singapore: Singapore University Press.

Former Australian Government Official I. 2019. "Former Lawyer, Attorney-General's Department, Australian Government" interviewed by Andreas Østhagen, Canberra, January 2019.

Forsberg, Tuomas. 1996. "Explaining Territorial Disputes: From Power Politics to Normative Reasons." *Journal of Peace Research* 33 (4): 433–49.

Foucault, Michel. 1972. *The Archaeology of Knowledge*. New York: Pantheon.

Franck, Thomas M. 1995. *Fairness in International Law and Institutions*. New York: Oxford University Press.

Franck, Thomas M., and Dennis M. Sughrue. 1993. "The International Role of Equity-as-Fairness." *The Georgetown Law Journal* 81: 563–595.

Frazier, D., and R. Stewart-Ingersoll. 2010. "Regional Powers and Security: A Framework for Understanding Order within Regional Security Complexes." *European Journal of International Relations* 16 (4): 731–53. https://doi.org/10.1177/1354066109359847.

Friedheim, Robert L. 1993. *Negotiating the New Ocean Regime*. Columbia: University of South Carolina Press.

Furlong, Paul, and David March. 2002. "A Skin Not a Sweater: Ontology and Epistemology in Political Science." In *Theory and Methods in Political Science*, edited by David March and Gerry Stoker, 2nd ed., 17–41. London: Palgrave Macmillan.

Fyfe, Nigel, and Greg French. 2005. "Australia–New Zealand." In *International Maritime Boundaries, Vol. 5*, edited by David A. Colson and Robert W. Smith, 3759–66. Dordrecht: Martinus Nijhoff.

Galvis, Richardo Abello, and Walter Arévalo. 2018. "Colombia's Maritime Boundaries" interviewed by Andreas Østhagen, Bogota, May 2018.

Gänsbauer, Anja, Ulrike Bechtold, and Harald Wilfing. 2016. "SoFISHticated Policy – Social Perspectives on the Fish Conflict in the Northeast Atlantic." *Marine Policy* 66: 93–103. https://doi.org/10.1016/j.marpol.2016.01.014.

Gaynor, Andrea. 2013. "Environmental Transformations." In *The Cambridge History of Australia: Volume 1: Indigenous and Colonial Australia*, edited by Alison Bashford and Stuart Macintyre, 269–93. Port Melbourne: Cambridge University Press.

Gibbs, Walter. 2010. "Russia and Norway Reach Accord on Barents Sea." *New York Times Europe*. http://www.nytimes.com/2010/04/28/world/europe/28norway.html?_r=1.

Gilpin, Robert. 2001. *Global Political Economy: Understanding the International Economic Order*. Princeton: Princeton University Press.

Global Affairs Canada. 2018. "Canada and the Kingdom of Denmark (with Greenland) Announce the Establishment of a Joint Task Force on Boundary Issues." Government of Canada, 2018.

Goertz, Gary, and Paul F. Diehl. 1992. *Territorial Changes and International Conflict.* 1st ed. New York: Routledge.

Goldsmith, Jack L., and Eric A. Posner. 2005. *The Limits of International Law.* 1st ed. Vol. 1. New York: Oxford University Press.

Goldstein, Judith, Miles Kahler, Robert O. Keohane, and Anne-Marie Slaughter. 2000. "Introduction: Legalization and World Politics." *International Organization* 54 (3): 385–99.

Goldstein, Judith, and Robert O. Keohane. 1993. "Ideas and Foreign Policy: An Analytical Framework." In *Ideas and Foreign Policy: Beliefs, Institutions and Political Change*, edited by Judith Goldstein and Robert O. Keohane, 3–30. Ithaca: Cornell University Press.

Gourevitch, Peter. 1978. "The Second Image Reversed: The International Sources of Domestic Politics." *International Organization* 32 (4): 881–912.

Government of Canada. 1970. *Act to Amend the Territorial Sea and Fishing Zones Act, SC 1969–70.*

Government of Colombia. 1978. *Act No. 10 of 4 August 1978 Establishing Rules Concerning the Territorial Sea, the Exclusive Economic Zone and the Continental Shelf and Regulating Other Matters.*

Gray, David H. 1997. "Canada's Unresolved Maritime Boundaries." *IBRU Boundary and Security Bulletin* 5 (3): 61.

Great Britain–Russia. 1825. "Great Britain–Russia: Limits of Their Respective Possessions on the North-West Coast of America and the Navigation of the Pacific Ocean." 75 CTS 95: 16 February 1825.

Great Britain–United States (1846). Treaty Establishing the Boundary in the Territory on the Northwest Coast of America Lying Westward of the Rocky Mountains. https://avalon.law.yale.edu/19th_century/br-1846.asp

Green, L. C. 1952. "The Anglo-Norwegian Fisheries Case, 1951 (I. C. J. Reports 1951, p. 116)." *The Modern Law Review* 15 (3): 373–77.

Griffiths, Franklyn, Rob Huebert, and Whitney P. Lackenbauer. 2011. *Canada and the Changing Arctic: Sovereignty, Security and Stewardship.* Waterloo: Wilfrid Laurier University Press.

Griffiths, Sian. 2010. "US–Canada Arctic Border Dispute Key to Maritime Riches." *BBC News*, August 2, 2010. https://www.bbc.com/news/world-us-canada-10834006.

Guggenheim, Lana B. 2019. "Tensions Mount in the Eastern Mediterranean as Turkey Seeks to Drill in Cyprus' EEZ." *SouthEUSummit*, May 16, 2019. https://www.southeusummit.com/europe/cyprus/tensions-mount-in-the-eastern-mediterranean-as-turkey-seeks-to-drill-in-cyprus-eez/.

Gullett, Warwick, and Clive Schofield. 2007. "Pushing the Limits of the Law of the Sea Convention: Australian and French Cooperative Surveillance and Enforcement in the Southern Ocean." *International Journal of Marine and Coastal Law* 22 (4): 545–83.

Hafner-Burton, Emilie M., David G. Victor, and Yonatan Lupu. 2012. "Political Science Research on International Law: The State of the Field." *The American Journal of International Law* 106 (1): 47–97.

Hannigan, John. 2017. "Toward a Sociology of Oceans." *Canadian Review of Sociology* 54 (1): 8–27.

Haraldstad, Marie. 2014. "Embetsverkets rolle i utformingen av norsk sikkerhetspolitikk: Nærområdeinitiativet (The Role of the Bureaucracy in the Shaping of Norway's Security Policy: The Close Area Initiative)." *Internasjonal Politikk* 72 (4): 431–51.

Harrabin, Roger. 2017. "Ocean Plastic a 'Planetary Crisis' – UN." *BBC News*, December 5, 2017.

Harris, David, and Sandesh Sivakumaran. 2015. *Cases and Materials on International Law*. 8th ed. London: Sweet & Maxwell.

Harrison, James. 2011. *Making the Law of the Sea: A Study in the Development of International Law*. Cambridge: Cambridge University Press.

Harsson, Bjørn Geirr, and George Preiss. 2012. "Norwegian Baselines, Maritime Boundaries and the UN Convention on the Law of the Sea." *Arctic Review on Law and Politics* 3 (1): 108–29.

Hasan, Md. Monjur, and He Jian. 2019. "Protracted Maritime Boundary Dispute Resolutions in the Bay of Bengal: Issues and Impacts." *Thalassas: An International Journal of Marine Sciences* 35 (1): 323–40. https://doi.org/10.1007/s41208-019-0126-1.

Henriksen, Tore, and Geir Ulfstein. 2011. "Maritime Delimitation in the Arctic: The Barents Sea Treaty." *Ocean Development & International Law* 42 (1–2): 1–21.

Hensel, Paul R. 1999. "Charting a Course to Conflict: Territorial Issues and Interstate Conflict, 1816–1992." In *The Road Map to War*, edited by Paul F. Diehl, 115–46. Nashville: Vanderbilt University Press.

Hensel, Paul R., S. McLaughlin Mitchell, Thomas E. Sowers II, and Clayton L. Thyne. 2008. "Bones of Contention: Comparing Territorial, Maritime and River Issues." *Journal of Conflict Resolution* 52 (1): 117–43.

Hirano, Mutsumi. 2014. "The Maritime Dispute in Sino–Japanese Relations: Domestic Dimensions." *Asian Perspective* 38: 263–84. http://journals.rienner.com/doi/abs/10.5555/0258-9184-38.2.263.

Hoel, Alf Håkon. 2009. "Do We Need a New Legal Regime for the Arctic Ocean?" *The International Journal of Marine and Coastal Law* 24 (2): 443–56.

Hoel, Alf Håkon. 2014. "The Legal-Political Regime in the Arctic." In *Geopolitics and Security in the Arctic*, edited by Rolf Tamnes and Kristine Offerdal, 49–72. New York: Routledge.

Hollis, Martin, and Steve Smith. 1996. "A Response: Why Epistemology Matters in International Theory." *Review of International Studies* 22 (1): 111–16.

Holsbø, Ida. 2011. "Hvordan var det dulig å komme frem til en endelig løsning i delelinjespørsmålet mellom Norge og Russland i Barentshavet i 2010, og ikke i 1978? (How Was It Possible to Find a Final Solution to the Boundary Dispute between Norway and Russia?)." University of Tromsø.

Holsti, Kalevi J. 1991. *Peace and War: Armed Conflicts and International Order, 1648–1989*. Cambridge: Cambridge University Press.

Holtsmark, Sven G, ed. 2015. *Naboer i frykt og forventning: Norge og Russland 1917–2014 (Neighbours in Fear and Anticipation: Norway and Russia 1917–2014)*. Oslo: Pax Forlag.

Hong, Nong. 2010. "Law and Politics in the South China Sea Assessing the Role of UNCLOS in Ocean Dispute Settlement." *ProQuest Dissertations and Theses*. https://doi.org/10.4324/9780203111215.

Hønneland, Geir. 1999. "Co-Management and Communities in the Barents Sea Fisheries." *Human Organization* 58 (4): 397–404.

Hønneland, Geir. 2012. *Making Fishery Agreements Work: Post-Agreement Bargaining in the Barents Sea*. Cheltenham, UK and Northampton, MA, USA: Edward Elgar Publishing.

Hønneland, Geir. 2013. *Hvordan skal Putin ta Barentshavet tilbake? (How Shall Putin Reclaim the Barents Sea?)*. Bergen: Fagbokforlaget.

Hønneland, Geir. 2016. *Russia and the Arctic: Environment, Identity and Foreign Policy*. London: I. B. Tauris.

Hønneland, Geir, and Anne-Kristin Jørgensen. 2015. "Kompromisskulturen i Barentshavet (the Culture of Compromise in the Barents Sea)." In *Norge Og Russland: Sikkerhetspolitiske Utfordringer i Nordområdene (Norway and Russia: Security Challenges in the High North)*, edited by Tormod Heier and Anders Kjølberg, 57–68 Oslo: Universitetsforlaget.

Hopf, Ted. 2002. *Social Construction of Foreign Policy: Identities and Foreign Policies, Moscow, 1955 and 1999*. Ithaca: Cornell University Press.

Howlett, Michael, M. Ramesh, and Anthony Perl. 2009. *Studying Public Policy: Policy Cycles & Policy Subsystems*. 3rd ed. Oxford: Oxford University Press.

Hui, Wang. 2014. *China from Empire to Nation-State*. Cambridge: Harvard University Press.

Hurd, Ian. 1999. "Legitimacy and Authority in International Politics." *International Organization* 53 (2): 379–408.

Huth, Paul K. 1998. *Standing Your Ground: Territorial Disputes and International Conflict*. Ann Arbor: University of Michigan Press.

Huth, Paul K., Sarah E. Croco, and Benjamin J. Appel. 2011. "Does International Law Promote the Peaceful Settlement of International Disputes? Evidence from the Study of Territorial Conflicts since 1945." *American Political Science Review* 105 (2): 415–36.

Iceland–Norway. 2008. *Agreement between Iceland and Norway Concerning Transboundary Hydrocarbon Deposits, 3 November 2008*. http://www.nea.is/media/olia/JM_unitisation_agreement_Iceland_Norway_2008.pdf.

ICJ. 1969. *North Sea Continental Shelf Cases*. http://www.icj-cij.org/docket/files/51/5535.pdf.

ICJ. 1984. *Delimitation of the Maritime Boundary in the Gulf of Maine Area (Canada/United States of America)*.

ICJ. 1993. *Maritime Delimitation in the Area between Greenland and Jan Mayen (Denmark v Norway)*. http://www.icj-cij.org/docket/files/78/6743.pdf.

ICJ. 2009. *Maritime Delimitation in the Black Sea (Romania v. Ukraine)*. International Court of Justice. http://www.icj-cij.org/docket/files/132/14987.pdf.

ICJ. 2012. *No Title*. https://www.icj-cij.org/files/case-related/124/124-20121119-JUD-01-00-EN.pdf.

ICJ. 2018. "Territorial and Maritime Dispute (Nicaragua v. Colombia): Overview of the Case." https://www.icj-cij.org/en/case/124.

International Energy Agency (IEA). 2017. "World Energy Outlook 2017." Paris. https://www.iea.org/weo2017/.

Inuvialuit Regional Corporation. 1984. *Inuvialuit Final Agreement (as Amended)*. https://www.inuvialuitland.com/resources/Inuvialuit_Final_Agreement.pdf.

IPCC. 2013. "Summary for Policymakers." In *Climate Change 2013: The Physical Science Basis. Contribution of Working Group I to the Fifth Assessment Report of the Intergovernmental Panel on Climate Change*, edited by T.F. Stocker, D. Qin, G.-K. Plattner, M. Tignor, S.K. Allen, J. Boschung, A. Nauels, Y. Xia, V. Bex, and P.M. Midgley, 2013 ed., 1–29. Cambridge: Cambridge University Press.

Irwin, Paul C. 1980. "Settlement of Maritime Boundary Disputes: An Analysis of the Law of the Sea Negotiations." *Ocean Development & International Law* 8 (2): 105–48.

Jackson, Robert H. 1987. "Quasi-States, Dual Regimes, and Neoclassical Theory: International Jurisprudence and the Third World." *International Organization* 41 (4): 519–49.

Jaeckel, Aline. 2016. "Deep Seabed Mining and Adaptive Management: The Procedural Challenges for the International Seabed Authority." *Marine Policy* 70: 205–11.

Jaeckel, Aline, Kristina M. Gjerde, and Jeff A. Ardron. 2017. "Conserving the Common Heritage of Humankind – Options for the Deep-Seabed Mining Regime." *Marine Policy* 78 (November 1994): 150–57.

Jagota, S. P. 1985. *Maritime Boundary*. Dordrecht: Martinus Nijhoff.

Jensen, Leif Christian. 2014. "The Times They Are A-Changin': Norsk sikkerhet og usikkerhet i Nordområdene." *Internasjonal Politikk* 72 (1): 7–29.

Jensen, Leif Christian. 2017. "An Arctic 'Marriage of Inconvenience': Norway and the Othering of Russia." *Polar Geography* 40 (2): 121–43. https://doi.org/10.1080/1088937X.2017.1308975.

Jensen, Øystein. 2014a. *Noreg og havets folkerett (Norway and the Law of the Sea)*. Trondheim: Akademia Forlag.

Jensen, Øystein. 2014b. "The Commission on the Limits of the Continental Shelf: An Administrative, Scientific, or Judicial Institution?" *Ocean Development & International Law* 45 (2): 171–85. https://doi.org/10.1080/00908320.2014.898921.

Jervis, Robert. 1978. "Cooperation under the Security Dilemma." *World Politics* 30 (2): 167–214.

Jianfei, Liu. 2012. "Land and Maritime Boundary Disputes' Challenges over China's Rise." *Contemporary International Relations* 22 (5): 45–48.

Jockel, Joseph T., and Joel J. Sokolsky. 2009. "Canada and NATO: Keeping Ottawa in, Expenses down, Criticism out … and the Country Secure." *International Journal* 64 (2): 315–36.

Jockel, Joseph T., and Joel J. Sokolsky. 2012. "Continental Defence: 'Like Farmers Whose Lands Have a Common Concession Line.'" In *Canada's National Security in the Post-9/11 World*, edited by David McDonough, 114–36. Toronto: University of Toronto Press.

Jóhannesson, Guðni Th. 2013. "The Jan Mayen Dispute between Iceland and Norway, 1979–1981: A Study in Successful Diplomacy?" Tromsø. http://gudnith.is/efni/jan_mayen_dispute_24_jan_2013.

Johnston, A. I. 2001. "Treating, International Institutions as Social Environments." *International Studies Quarterly* 45 (4): 487–515.

Johnston, Douglas M. 1988. *The Theory and History of Ocean Boundary-Making*. Montreal: McGill-Queen's University Press.

Johnston, Douglas M., and Phillip M. Saunders. 1988. "Ocean Boundary Issues and Developments in Regional Perspective." In *Ocean Boundary Making: Regional Issues and Developments*, edited by Douglas M. Johnston and Phillip M. Saunders, 313–49. London: Croom Helm.

Johnston, Douglas M., and Mark J. Valencia. 1991. *Pacific Ocean Boundary Problems: Status and Solutions*. Dordrecht: Martinus Nijhoff.

Jørgensen, Anne-Kristin, and Andreas Østhagen. 2020. "Norges vern av suverene rettigheter rundt Svalbard: Russiske persepsjoner og reaksjoner." *Internasjonal Politikk* 78 (2): 167–92. https://doi.org/10.23865/intpol.v78.1838.

Junhong, Liu. 2012. "China's Maritime Frontiers and the Global System." *Contemporary International Relations* 22 (5): 77–81.

Kaarsted, Tage. 1992. *De danske ministerier 1953–1972*. Odense: Universitetsforlaget.

Kaczynski, Vlad M. 2007. "The Kuril Islands Dispute Between Russia and Japan: Perspectives of Three Ocean Powers." *Russian Analytical Digest* 20 (May 1, 2007): 6–8.

Kaplan, Robert D. 2011. "The South China Sea Is the Future of Conflict." *Foreign Policy*, 188: 76–85.

Karlsbakk, Jonas. 2008. "Norway and Iceland Sign Border Treaty." *Barents Observer*, November 5, 2008. http://barentsobserver.com/en/node/20950.

Katzenstein, Peter J., and Rudra Sil. 2008. "Eclectic Theorizing in the Study and Practice of International Relations." In *The Oxford Handbook of International Relations*, edited by Christian Reus-Smit and Duncan Snidal, 109–28. Oxford: Oxford University Press.

Kaye, Stuart. 1994. "The Torres Strait Islands: Constitutional and Sovereignty Questions Post-Mabo." *University of Queensland Law Journal* 18 (1): 38–49.

Kaye, Stuart. 2001. *Australia's Maritime Boundaries*. 2nd ed. University of Wollongong.

Keck, Margaret E., and Kathryn Sikkink. 1998. *Activists Beyond Borders: Advocacy Networks in International Politics*. Ithaca: Cornell University Press.

Kelly, Robert E. 2007. "Security Theory in the 'New Regionalism.'" *International Studies Review* 9 (2): 197–229.

Keohane, Robert O. 1984. *After Hegemony: Cooperation and Discord in the World Political Economy*. Princeton: Princeton University Press.

Keohane, Robert O., and Joseph S. Nye. 2012[1977]. *Power and Interdependence*. 4th-Kindle ed. Boston: Longman.

Kindingstad, Torbjørn. 2002. *Norges Oljehistorie*. Oslo: Wigestrand.

King, Anthony. 1973. "Ideas, Institutions and the Policies of Governments: A Comparative Analysis: Part III." *British Journal of Political Science* 3 (4): 409–23.

Kingdon, John W. 1984. *Agendas, Alternatives and Public Policies*. Boston: Little, Brown.

Kirkey, Christopher. 1995. "Delineating Maritime Boundaries: The 1977–78 Canada–U.S. Beaufort Sea Continental Shelf Delimitation Boundary Negotiations." *Canadian Review of American Studies* 25 (2): 49–66.

Klein, Natalie. 2006. "Provisional Measures and Provisional Arrangements in Maritime Boundary Disputes." *International Journal of Marine and Coastal Law* 21 (4): 423–60. https://doi.org/10.1163/157180806779441129.

Kleine-Ahlbrandt, Stephanie. 2012. "Fish Story: The Risk of Conflict in the South China Sea is Real. But Not for the Reasons You Might Think." *Foreign Policy*, 2012. http://foreignpolicy.com/2012/06/25/fish-story/.

Kleinsteiber, Meghan. 2013. "Nationalism and Domestic Politics as Drivers of Maritime Conflict." *SAIS Review of International Affairs* 33 (2): 15–19.

Kolb, Robert. 2003. *Case Law on Equitable Maritime Delimitation, Digest and Commentaries*. The Hague: Martinus Nijhoff.

Kraska, James. 2011. *Maritime Power and the Law of the Sea: Expeditionary Operations in World Politics*. Oxford: Oxford University Press.

Krasner, Stephen D., ed. 1983. *International Regimes*. Ithaca: Cornell University Press.

Krasner, Stephen D. 1999. *Sovereignty: Organized Hypocrisy*. Princeton: Princeton University Press.

Kratochwil, Friedrich. 1986. "Of Systems, Boundaries, and Territoriality: An Inquiry into the Formation of the State System." *World Politics* 39 (1): 27–52.

Kwiatkowska, Barbara. 2004. "Economic and Environmental Considerations in Maritime Boundary Delimitations." In *International Maritime Boundaries, Volumes*

2–3, edited by Jonathan I. Charney and Lewis M. Alexander, 75–114. The Hague: Martinus Nijhoff.

Kyodo. 2017. "Japan Successfully Undertakes Large-Scale Deep-Sea Mineral Extraction." *Japantimes*, September 26, 2017. https://www.japantimes.co.jp/news/2017/09/26/national/japan-successfully-undertakes-large-scale-deep-sea-mineral-extraction/#.WlXV5qhKujt.

Lackenbauer, Whitney P. 2013. "Harper's Arctic Evolution." *The Globe and Mail*, 2013. http://www.theglobeandmail.com/globe-debate/harpers-arctic-evolution/article13852195/.

Laffoley, D., and J. M. Baxter. 2016. "Explaining Ocean Warming: Causes, Scale, Effects and Consequences." Gland: International Union for Conservation of Nature and Natural Resources. https://doi.org/10.2305/IUCN.CH.2016.08.en.

Lalonde, Suzanne. 2002. *Determining Boundaries in a Conflicted World: The Role of Uti Possidetis*. Montreal: McGill-Queen's University Press.

Lalonde, Suzanne. 2010. "A Network of Marine Protected Areas in the Arctic: Promises and Challenges." In *Changes in the Arctic Environment and the Law of the Sea*, edited by Myron Nordquist, John Norton Moore, and Tomas H. Heidar, 131–42. Leiden: Brill Nijhoff.

Lang, Anthony F, and Amanda Russell Beattie. 2009. "War, Torture and Terrorism." *International Relations*. https://doi.org/10.4324/9780203888452.

LaRosa, Michael J., and Germán R. Mejía. 2012. *Colombia: A Concise Contemporary History*. Plymouth: Rowman & Littlefield.

Lasswell, Harold D. 1936. "Politics Who Gets What, When, How." McGraw-Hill. http://www.policysciences.org/classics/politics.pdf.

Lavrov, Sergei, and Jonas Gahr Støre. 2010. "Canada, Take Note: Here's How to Resolve Maritime Disputes." *The Globe and Mail*, September 21, 2010. http://www.theglobeandmail.com/opinion/canada-take-note-heres-how-to-resolve-maritime-disputes/article4326372/.

Lejeusne, Christophe, Pierre Chevaldonné, Christine Pergent-Martini, Charles F. Boudouresque, and Thierry Pérez. 2010. "Climate Change Effects on a Miniature Ocean: The Highly Diverse, Highly Impacted Mediterranean Sea." *Trends in Ecology and Evolution*. https://doi.org/10.1016/j.tree.2009.10.009.

Levin, Lisa A., and Nadine Le Bris. 2015. "The Deep Ocean under Climate Change." *Science* 350 (6262): 766–68. https://doi.org/10.1126/science.aad0126.

Levin, Lisa A., Kathryn Mengerink, Kristina M. Gjerde, Ashley A. Rowden, Cindy Lee Van Dover, Malcolm R. Clark, Eva Ramirez-Llodra et al. 2016. "Defining 'Serious Harm' to the Marine Environment in the Context of Deep-Seabed Mining." *Marine Policy* 74 (September): 245–59. https://doi.org/10.1016/j.marpol.2016.09.032.

Levy, Marc A., Robert O. Keohane, and Peter M. Haas. 1993. "Improving the Effectiveness of International Environmental Institutions." In *Institutions for the Earth: Sources of Effective International Environmental Protection*, edited by Peter M. Haas, Robert O. Keohane, and Marc A. Levy, 397–426. Cambridge: MIT Press.

Levy, Marc A., Oran R. Young, and Michael Zürn. 1995. "The Study of International Regimes." *European Journal of International Relations* 1 (3): 267–330.

Lichfield, John. 2018. "The Brexiteers' 'Take Back Our Waters' Pledge Is Meaningless Hype." *The Guardian*, November 24, 2018. https://www.theguardian.com/commentisfree/2018/nov/23/brexit-waters-fishing-industry-eu.

Lijphart, Arend. 1989. "Democratic Political Systems." *Journal of Theoretical Politics* 1 (1): 33–48.

Linderfalk, Ulf. 2016. "The Jan Mayen Case (Iceland/Norway): An Example of Successful Conciliation." *SSRN*, no. May 24, 2016: 1–26. https://ssrn.com/abstract =2783622.

Livingston, Steven G. 1992. "The Politics of International Agenda-Setting: Reagan and North-South Relations." *International Studies Quarterly* 36: 313–30.

Londoño, Julio Paredes. 2015. *Colombia En El Laberinto Del Caribe (Colombia in the Caribbean Labyrinth)*. Bogota DC: Universidad del Rosario.

Londoño, Julio Paredes. 2018. "Former Foreign Minister of Colombia, Colombian Government" interviewed by Andreas Østhagen, Bogota, May 2018.

Lovelock, James. 1979. *Gaia: A New Look at Life on Earth*. Oxford: Oxford University Press.

Lowe, David. 2013. "Security." In *The Cambridge History of Australia: Volume 2 – The Commonwealth of Australia*, edited by Alison Bashford and Stuart Macintyre, 494–517. Port Melbourne: Cambridge University Press.

Lusthaus, Jonathan. 2010. "Shifting Sands: Sea Level Rise, Maritime Boundaries and Inter-State Conflict." *Politics* 30 (2): 113–18. https://doi.org/10.1111/j.1467-9256 .2010.01374.x.

Mackrael, Kim. 2012. "Canada, Denmark Closer to Settling Border Dispute." *Globe and Mail*, November 29, 2012.

Mahan, Alfred T., and Charles Beresford. 1894. "Possibilities of an Anglo-American Reunion." *The North American Review* 159 (456): 551–73.

Maier, Charles S. 2016. *Once Within Borders: Territories of Power, Wealth, and Belonging since 1500*. Cambridge: Harvard University Press.

Manheim, Jarol B., and Robert B. Albritton. 1984. "Changing National Images: International Public Relations and Media Agenda Setting." *The American Political Science Review* 78 (3): 641–57.

Maritime Delimitation in the Area between Greenland and Jan Mayen (Denmark v. Norway). 1993.

Mazmanian, Daniel A., and Paul A. Sabatier. 1980. "A Multivariate Model of Public Policy-Making." *American Journal of Political Science* 24 (3): 439–68.

McCreery, Cindy, and Kirsten McKenzie. 2013. "The Australian Colonies in a Maritime World." In *The Cambridge History of Australia: Volume 1: Indigenous and Colonial Australia*, edited by Alison Bashford and Stuart Macintyre, 560–84. Port Melbourne: Cambridge University Press.

McDorman, Ted L. 2002. "The Role of the Commision on the Limits of Continental Shelf: A Technical Body in a Political World." *International Journal of Marine and Coastal Law* 17: 301.

McDorman, Ted L. 2009. *Salt Water Neighbors: International Ocean Law Relations between the United States and Canada*. New York: Oxford University Press.

McHarg, Aileen, Barry Barton, Adrian Bradbrook, and Lee Godden. 2010. *Property and the Law in Energy and Natural Resources*. New York: Oxford University Press.

McRae, Donald M. 1989. "Canada and the Delimitation of Maritime Boundaries." In *Canadian Oceans Policy: National Strategies and the New Law of the Sea*, edited by Donald M. McRae and Gordon Munro, 145–64. Vancouver: UBC Press.

Mearsheimer, John J. 1995. "The False Promise of International Institutions." *International Security* 19 (3): 5–49.

Mearsheimer, John J. 2001. *The Tragedy of Great Power Politics*. New York: W. W. Norton & Company.

Mière, Christian Le. 2011. "The Return of Gunboat Diplomacy." *Survival* 53 (5): 53–68.

Milner, Helen V. 1998. "Rationalizing Politics: The Emerging Synthesis of International, American, and Comparative Politics." *International Organization* 52 (4): 759–86. https://doi.org/10.1162/002081898550743.

Mishra, Rahul. 2017. "Code of Conduct in the South China Sea: More Discord than Accord." *Maritime Affairs* 13 (2): 62–75.

Moe, Arild, Daniel Fjærtoft, and Indra Øverland. 2011. "Space and Timing: Why Was the Barents Sea Delimitation Dispute Resolved in 2010?" *Polar Geography* 34 (3): 145–62.

Molenaar, Erik J. 2012. "Fisheries Regulation in the Maritime Zones of Svalbard." *The International Journal of Marine and Coastal Law* 27 (1): 3–58.

Moravcsik, Andrew. 1997. "Taking Preferences Seriously: A Liberal Theory of International Politics." *International Organization* 51 (4): 229.

Nasu, Hitoshi, and Donald R. Rothwell. 2014. "Re-Evaluating the Role of International Law in Territorial and Maritime Disputes in East Asia." *Asian Journal of International Law* 4 (1): 55–79.

Nelson, L. D. M. 1990. "The Roles of Equity in the Delimitation of Maritime Boundaries." *The American Journal of International Law* 84 (4): 837–58.

Nemeth, Stephen C., Sara McLaughlin Mitchell, Elizabeth A. Nyman, and Paul R. Hensel. 2014. "Ruling the Sea: Managing Maritime Conflicts through UNCLOS and Exclusive Economic Zones." *International Interactions* 40 (5): 711–36.

Neumann, Iver B. 1996. "Self and Other in International Relations." *European Journal of International Relations* 2 (2): 139–74.

Neumann, Iver B, Walter Carlsnaes, John Kristen Skogan, Nina Græger, Pernille Rieker, Kristin M Haugevik, and Stina Torjesen. 2008. "Norge og alliansene: Gamle tradis-joner, nytt spillerom." Oslo: Norwegian Institute of International Affairs (NUPI).

Neumann, Iver B., and Sieglinde Gstöhl. 2006. "Lilliputians in Gulliver's World? Small States in International Relations." In *Annual ISA Conference*. San Diego: ISA.

Neumann, Thilo. 2010. "Norway and Russia Agree on Maritime Boundary in the Barents Sea and the Arctic Ocean." *The American Society of International Law Insights* 14 (34): https://www.asil.org/insights/volume/14/issue/34/norway-and-russia-agree-maritime-boundary-barents-sea-and-arctic-ocean

Nieto-Navia, Rafael. 2015. "Some Remarks on the Territorial and Maritime Dispute (Nicaragua v. Colombia) Case." In *Law of the Sea, From Grotius to the International Tribunal for the Law of the Sea*, edited by Lilian Castillo, 545–62. Leiden: Brill Nijhoff.

Nordquist, Myron H., and John Norton Moore. 2012. *Maritime Border Diplomacy*. Dordrecht: Martinus Nijhoff.

Norway–Iceland. 1981. *Agreement on the Continental Shelf between Iceland and Jan Mayen*. http://www.un.org/depts/los/LEGISLATIONANDTREATIES/PDFFILES/TREATIES/ISL-NOR1981CS.PDF.

Norway–UK (1965). Agreement relating to the delimitation of the continental shelf between the two countries. https://treaties.un.org/doc/Publication/UNTS/Volume%20551/volume-551-I-8043-English.pdf

Norwegian Diplomat I. 2017. "Former Lead Negotiator for Norway, Norwegian Ministry of Foreign Affairs, Norwegian Government" interviewed by Andreas Østhagen, Paris, January 2017.

Norwegian Government. 2007. "Agreement Signed between Norway and Russia on Maritime Delimitation in the Varangerfjord Area." Press Release, 2007. https://www.regjeringen.no/en/aktuelt/Agreement-signed-between-Norway-and-Russ/id476347/.

Norwegian Government. 2010. "Joint Statement on Maritime Delimitation and Cooperation in the Barents Sea and the Arctic Ocean." Press Release, 2010.

Norwegian Petroleum Directorate. 2017. "Norway's Petroleum History." *Norwegian Petroleum*, 2017. http://www.norskpetroleum.no/en/framework/norways-petroleum -history/.

Nweihed, Kaldone G. 1980. "EZ (Uneasy) Delimitation in the Semi-Enclosed Caribbean Sea: Recent Agreements between Venezuela and Her Neighbors." *Ocean Development & International Law* 8 (1): 1–33. https://doi.org/10.1080/00908328009545642.

Nweihed, Kaldone G. 1983. "El Caribe de La Pesca; Estudio Acerca de Las Pesquerias Del Caribe y Areas Adiacentes: Aspectos Economico, Social, Politico y Juridico" 1–2.

Nweihed, Kaldone G. 1993a. "Colombia–Costa Rica." In *International Maritime Boundaries, Vol. 1*, edited by Jonathan I. Charney and Lewis M. Alexander, 463–72. Dordrecht: Martinus Nijhoff.

Nweihed, Kaldone G. 1993b. "Colombia–Dominican Republic." In *International Maritime Boundaries, Vol. 1*, edited by Jonathan I. Charney and Lewis M. Alexander, 477–86. Dordrecht: Martinus Nijhoff.

Nweihed, Kaldone G. 1993c. "Colombia–Haiti." In *International Maritime Boundaries, Vol. 1*, edited by Jonathan I. Charney and Lewis M. Alexander, 491–98. Dordrecht: Martinus Nijhoff.

Nweihed, Kaldone G. 1993d. "Colombia–Honduras." In *International Maritime Boundaries, Vol. 1*, edited by Jonathan I. Charney and Lewis M. Alexander, 503–15. Dordrecht: Martinus Nijhoff.

Nweihed, Kaldone G. 1993e. "Colombia–Panama." In *International Maritime Boundaries Vol. 1*, edited by Jonathan I. Charney and Lewis M. Alexander, 519–30. Dordrecht: Martinus Nijhoff.

Nweihed, Kaldone G. 1993f. "Middle American and Caribbean Maritime Boundaries." In *International Maritime Boundaries Vol. 1*, edited by Jonathan I. Charney and Lewis M. Alexander, 271–83. Dordrecht: Martinus Nijhoff.

Nweihed, Kaldone G. 1998. "Colombia–Jamaica." In *International Maritime Boundaries, Vol. 3*, edited by Jonathan I. Charney and Lewis M. Alexander, 2179–98. Dordrecht: Martinus Nijhoff.

Nyman, Elizabeth. 2013. "Oceans of Conflict: Determining Potential Areas of Maritime Disputes." *SAIS Review of International Affairs* 33 (2): 5–14. https://doi.org/10.1353/sais.2013.0025.

Nyman, Elizabeth. 2015. "Offshore Oil Development and Maritime Conflict in the 20th Century: A Statistical Analysis of International Trends." *Energy Research and Social Science* 6: 1–7. https://doi.org/10.1016/j.erss.2014.10.006.

Nyman, Elizabeth, and Rachel Tiller. 2020. "'Is There a Court that Rules Them All'? Ocean Disputes, Forum Shopping and the Future of Svalbard." *Marine Policy* 113 (March). https://doi.org/10.1016/j.marpol.2019.103742.

Okafor-Yarwood, Ifesinachi. 2015. "The Guinea–Bissau–Senegal Maritime Boundary Dispute." *Marine Policy* 61 (November): 284–90.

Okonkwo, Theodore. 2017. "Maritime Boundaries Delimitation and Dispute Resolution in Africa." *Beijing Law Review* 8 (1): 55–78. https://doi.org/10.4236/blr.2017.81005.

Olsen, John Andreas. 2017. "NATO and the North Atlantic: Revitalising Collective Defence." London. https://rusi.org/publication/whitehall-papers/nato-and-north -atlantic-revitalising-collective-defence.

Ong, David M. 2015. "A Bridge Too Far? Assessing the Prospects for International Environmental Law to Resolve the South China Sea Disputes." *International Journal on Minority and Group Rights* 22: 578–97.

Oosterveld, Willem, Stephan De Spiegeleire, and Tim Sweijs. 2015. *Pushing the Boundaries: Territorial Conflict in Today's World*. The Hague: The Hague Centre for Strategic Studies.

Orttung, Robert W., and Andreas Wenger. 2016. "Explaining Cooperation and Conflict in Marine Boundary Disputes Involving Energy Deposits." *Region: Regional Studies of Russia, Eastern Europe, and Central Asia* 5 (1): 75–96. https://doi.org/10.1353/reg.2016.0001.

Osgood, Robert E. 1976. "Military Implications of the New Ocean Politics." *The Adelphi Papers* 16 (122): 10–16.

Østhagen, Andreas. 2016. "High North, Low Politics Maritime Cooperation with Russia in the Arctic." *Arctic Review on Law and Politics* 7 (1): 83–100.

Østhagen, Andreas. 2018a. "Geopolitics and Security in the Arctic." In *Routledge Handbook of the Polar Regions*, edited by Mark Nuttall, Torben R. Christensen, and Martin Siegert, 348–56. Abingdon: Routledge.

Østhagen, Andreas. 2018b. "Managing Conflict at Sea: The Case of Norway and Russia in the Svalbard Zone." *Arctic Review on Law and Politics* 9: 100–123.

Østhagen, Andreas. 2020. "Maritime Boundary Disputes: What Are They and Why Do They Matter?" *Marine Policy* 120 (October): 104–18. https://doi.org/10.1016/j.marpol.2020.104118.

Østhagen, Andreas. 2021a. "Norway's Arctic Policy: Still High North, Low Tension?" *The Polar Journal* 11 (1).

Østhagen, Andreas. 2021b. "Troubled Seas? The Changing Politics of Maritime Boundary Disputes." *Ocean & Coastal Management* 205 (May 2021): 105535. https://doi.org/10.1016/j.ocecoaman.2021.105535.

Østhagen, Andreas, and Andreas Raspotnik. 2018. "Crab! How a Dispute over Snow Crab Became a Diplomatic Headache between Norway and the EU." *Marine Policy* 98 (December 2018): 58–64. https://doi.org/10.1016/j.marpol.2018.09.007.

Østhagen, Andreas, and Andreas Raspotnik. 2020. "The EU in Antarctica: An Emerging Area of Interest, or Playing to the (Environmental) Gallery?" *European Foreign Affairs Review* 25 (2): 239–260.

Østhagen, Andreas, and Clive H. Schofield. 2021. "An Ocean Apart? Maritime Boundary Agreements and Disputes in the Arctic Ocean." *The Polar Journal*. https://doi.org/10.1080/2154896X.2021.1978234.

Østhagen, Andreas, Greg L. Sharp, and Paal S. Hilde. 2018. "At Opposite Poles: Canada's and Norway's Approaches to Security in the Arctic." *The Polar Journal* 8 (1): 163–81.

Østreng, Willy, and Yngvild Prydz. 2007. "Delelinjen i Barentshavet: Planlagt samarbeid versus uforutsett konflikt? (The Barents Sea Boundary: Planned Cooperation versus Conflict?)." Oslo. https://www.stortinget.no/Global/pdf/Utredning/Perspektiv07_04.pdf.

Oude Elferink, Alex G. 2007. "Maritime Delimitation between Denmark/Greenland and Norway." *Ocean Development & International Law* 38 (4): 375–80.

Oude Elferink, Alex G. 2013. *The Delimitation of the Continental Shelf between Denmark, Germany and the Netherlands: Arguing Law, Practicing Politics?* Cambridge: Cambridge University Press.

Oude Elferink, Alex G. 2015. "International Law and Negotiated and Adjudicated Maritime Boundaries: A Complex Relationship." *German Yearbook of International Law* 48.

Oude Elferink, Alex G., Tore Henriksen, and Signe Veierud Busch. 2018a. "Conclusion: Taking Stock and Looking Ahead." In *Maritime Boundary Delimitation: The Case*

Law – Is It Consistent and Predictable?, edited by Alex G. Oude Elferink, Tore Henriksen, and Signe Veierud Busch, 376–402. Cambridge: Cambridge University Press.

Oude Elferink, Alex G., Tore Henriksen, and Signe Veierud Busch. eds. 2018b. *Maritime Boundary Delimitation: The Case Law – Is It Consistent and Predictable?* Cambridge: Cambridge University Press.

Oude Elferink, Alex G, and Donald R Rothwell. 2004. *Oceans Management in the 21st Century: Institutional Frameworks and Responses.* Publications on Ocean Development.

Oxman, Bernard H. 1995. "International Maritime Boundaries: Political, Strategic, and Historical Considerations." *University of Miami Inter-American Law Review* 26 (2): 243–95.

Paasi, Anssi, and Eeva Kaisa Prokkola. 2008. "Territorial Dynamics, Cross-Border Work and Everyday Life in the Finnish–Swedish Border Area." *Space and Polity* 12 (1): 13–29. https://doi.org/10.1080/13562570801969366.

Paine, Lincoln. 2013. *The Sea & Civilization: A Maritime History of the World.* New York: Vintage Books.

Palacios, Marco. 2006. *Between Legitimacy and Violence: A History of Colombia, 1875–2002.* Durham: Duke University Press.

Pardo, Arvid. 1968. "Address by Ambassador Arvid Pardo." *Proceedings of the American Society of International Law at Its Annual Meeting (1921–1969)* 62 (April 25–27): 216–29.

Park, Choon-ho. 1993a. "Australia–France (New Caledonia)." In *International Maritime Boundaries, Vol. 1–2*, edited by Jonathan I. Charney and Lewis M. Alexander, 905–9. Dordrecht: Martinus Nijhoff.

Park, Choon-ho. 1993b. "Australia–Papua New Guinea." In *International Maritime Boundaries, Vol. 1–2*, edited by Jonathan I. Charney and Lewis M. Alexander, 929–34. Dordrecht: Martinus Nijhoff.

Park, Choon-ho. 1993c. "Australia–Solomon Islands." In *International Maritime Boundaries Vol. 1–2*, edited by Jonathan I. Charney and Lewis E. Alexander, 977–81. Dordrecht: Martinus Nijhoff.

Pauly, Daniel, and Dirk Zeller, eds. 2016. *Global Atlas of Marine Fisheries: A Critical Appraisal of Catches and Ecosystem Impacts.* Washington DC: Island Press.

PCA. 2016. *The South China Sea Arbitration (The Republic of Philippines v. The People's Republic of China).*

Pedersen, Torbjørn. 2009. "Endringer i internasjonal Svalbard politikk (Changes in International Svalbard Policy)." *Internasjonal Politikk* 67 (1): 31–44.

Pedersen, Torbjørn. 2017. "The Politics of Presence: The Longyearbyen Dilemma." *Arctic Review on Law and Politics* 8: 95–108.

Pedersen, Torbjørn, and Tore Henriksen. 2009. "Svalbard's Maritime Zones: The End of Legal Uncertainty?" *The International Journal of Marine and Coastal Law* 24 (1): 141–61.

Permanent Court of Arbitration. 1909. "Maritime Boundary Norway–Sweden (Grisbådarna)." *The Hague Justice Portal.* http://www.haguejusticeportal.net/index.php?id=6130.

Pharand, Donat. 2007. "The Arctic Waters and the Northwest Passage: A Final Revisit." *Ocean Development and International Law* 38 (1–2): 3–69. https://doi.org/10.1080/00908320601071314.

Pinsky, Malin L., Gabriel Reygondeau, Richard Caddell, Juliano Palacios-Abrantes, Jessica Spijkers, and William W. L. Cheung. 2018. "Preparing Ocean Governance for Species on the Move." *Science* 360 (6394): 1189–91. https://doi.org/10.1126/science.aat2360.

Pomeroy, Robert, John Parks, Richard Pollnac, Tammy Campson, Emmanuel Genio, Cliff Marlessy, Elizabeth Holle et al. 2007. "Fish Wars: Conflict and Collaboration in Fisheries Management in Southeast Asia." *Marine Policy* 31 (6): 645–56. http://linkinghub.elsevier.com/retrieve/pii/S0308597X07000413.

Powell, Haywood Jefferson. 1993. "'Cardozo's Foot': The Chancellor's Conscience and Constructive Trusts." *Law and Contemporary Problems* 56 (3): 7–27.

Prescott, Victor. 1993a. "Australia–Indonesia (Fisheries)." In *International Maritime Boundaries Vol. 1–2*, edited by Jonathan I. Charney and Lewis M. Alexander, 1229–36. Dordrecht: Martinus Nijhoff.

Prescott, Victor. 1993b. "Australia–Indonesia (Timor Gap)." In *International Maritime Boundaries, Vol. 1–2*, edited by Jonathan I. Charney and Lewis M. Alexander, 1245–54. Dordrecht: Martinus Nijhoff.

Prescott, Victor. 1993c. "Australia (Heard/McDonald Lslands)–France (Kerguelen Islands)." In *International Maritime Boundaries, Vol. 1–2*, edited by Jonathan I. Charney and Lewis M. Alexander, 1185–90. Dordrecht: Martinus Nijhoff.

Prescott, Victor. 1993d. "Region VI: Indian Ocean and South East Asian Maritime Boundaries." In *International Maritime Boundaries, Vol. 1–2*, edited by Jonathan I. Charney and Lewis M. Alexander, 305–11. Dordrecht: Martinus Nijhoff.

Prescott, Victor. 2002. "Australia–Indonesia." In *International Maritime Boundaries, Vol. 4*, edited by Jonathan I. Charney and Robert W. Smith, 2697–712. Dordrecht: Martinus Nijhoff.

Prescott, Victor, and Clive H. Schofield. 2005. *Maritime Political Boundaries of the World*. Leiden: Martinus Nijhoff Publishers.

Prescott, Victor, and G. Triggs. 2005. "Australia–East Timor." In *International Maritime Boundaries, Vol. 5*, edited by David A. Colson and Robert W. Smith, 3806–19. Dordrecht: Martinus Nijhoff.

Prip, Christian. 2017. "The Arctic Council and Biodiversity." In *Arctic Governance: Law and Politics. Volume 1*, edited by Svein Vigeland Rottem and Ida Folkestad Soltvedt, 205–30. London: I. B. Tauris.

Putnam, Robert D. 1988. "Diplomacy and Domestic Politics: The Logic of Two-Level Games." *International Organization* 42 (3): 427–60.

Qiu, Wenxian, and Warwick Gullett. 2017. "Quantitative Analysis for Maritime Delimitation: Reassessing the Bay of Bengal Delimitation between Bangladesh and Myanmar." *Marine Policy* 78 (March 2016): 45–54. https://doi.org/10.1016/j.marpol.2017.01.011.

Rattray, K. O., A. Kirton, and P. Robinson. 1974. "The Effects of the Existing Law of the Sea on the Development of the Caribbean Region and the Gulf of Mexico." In *Caribbean Study and Dialogue*, 251–75. Msida: Malta University Press.

Ravin, Mom. 2005. "Law of the Sea: Maritime Boundaries and Dispute Settlement Mechanisms." http://www.un.org/depts/los/nippon/unnff_programme_home/fellows_pages/fellows_papers/mom_0506_cambodia.pdf.

Ravna, Øivind. 2012. "Samerett og samiske rettigheter i Norge (Sami Law and Sami Rights in Norway)." In *Juss i Nord: Hav, fisk og urfolk: En hyllest til det juridiske fakultet ved Universitetet i Tromsøs 25-årsjubileum (Law in the North: Ocean, Fish and Indigenous Peoples: A Tribute to the Legal Faculty at the University of Tromsø's 25th Anniversary)*. Oslo: Gyldendal.

Rayfuse, Rosemary. 2009. "W(h)Ither Tuvalu? International Law and Disappearing States." *University of New South Wales Faculty of Law Research Series – Paper 9.* Sydney.

Renouf, J. K. 1988. "Canada's Unresolved Maritime Boundaries." *Geodesy and Gemoatics Engineering UNB* Technical (134).

Retzer, Berit Ruud. 2017. *Jens Evensen: Mannen som gjorde Norge større (Jens Evensen: The Man Who Made Norway Larger).* Oslo: Gyldendal.

Reus-Smit, Christian. 2004. *The Politics of International Law.* Cambridge: Cambridge University Press.

Reuters. 2016. "Ban on New Arctic Drilling Gives Canada Leg Up, U.S. Lawmakers Say." *Reuters*, December 21, 2016. https://www.reuters.com/article/us-canada-arctic-drilling-idUSKBN14A261.

Rico. 2018. "Ortega Accuses Colombia of Coup to Overthrow Him." *Today Nicaragua*, August 15, 2018. https://todaynicaragua.com/ortega-accuses-colombia-of-coup-to-overthrow-him/.

Risse, Thomas. 2000. "'Let's Argue!': Communicative Action in World Politics." *International Organization* 54 (1): 1–39.

Riste, Olav. 2005. *Norway's Foreign Relations: A History.* Oslo: Universitetsforlaget.

Roach, J. Ashley. 2014. "Today's Customary International Law of the Sea." *Ocean Development and International Law* 45 (3): 239–59.

Robinson, Piers. 1999. "The CNN Effect: Can the News Media Drive Foreign Policy?" *Review of International Studies* 25 (2): 301–9. https://doi.org/10.1017/S0260210599003010.

Robinson, Piers. 2008. "The Role of Media and Public Opinion." In *Foreign Policy: Theories, Actors, Cases,* edited by Steve Smith, Amelia Hadfield and Tim Dunne, 186–205. Oxford: Oxford University Press.

Rothwell, Donald. 1996. *The Polar Regions and the Development of International Law. Cambridge Studies in International and Comparative Law.* Cambridge: Cambridge University Press.

Rothwell, Donald R. 2012. "International Straits and Trans-Arctic Navigation." *Ocean Development and International Law* 43 (3): 267–82.

Rothwell, Donald R., and Tim Stephens. 2016. *The International Law of the Sea.* 2nd ed. Oxford: Bloomsbury.

Ruggie, John Gerard. 1993. "Territoriality and Beyond: Problematizing Modernity in International Relations." *International Organization* 47 (1): 139–74.

Russell, Denise. 2010. *Who Rules the Waves? Piracy, Overfishing and Mining the Ocean.* London: Pluto Press.

Russia–Norway. 2007. "Agreement between the Russian Federation and the Kingdom of Norway on the Maritime Delimitation in the Varangerfjord Area (2007)." *UN Law of the Sea Bulletin 42*, 2007.

Ryggvik, Helge. 2014. "Forhandlingene om Norges kontinentalsokkel (Negotiations over Norway's Continental Shelf)." Store Norske Leksikon. 2014. https://snl.no/Forhandlingene_om_Norges_kontinentalsokkel.

Sack, Robert. 1986. *Human Territoriality: Its Theory and History.* Cambridge: Cambridge University Press.

Salayo, N., K. Viswanathan, M. Ahmed, and L. Garces. 2006. "An Overview of Fisheries Conflicts in South and Southeast Asia: Recommendations, Challenges and Directions." *Naga The WorldFish Center Quarterly* 29 (1 & 2): 11–20. http://www.worldfishcenter.org/resource_centre/overview.pdf.

Sandner, Gerhard, and Beate Ratter. 1991. "Topographical Problem Areas in the Delimitation of Maritime Boundaries and Their Political Relevance: Case Studies from the Western Caribbean." *Ocean and Shoreline Management* 15 (4): 289–308. https://doi.org/10.1016/0951-8312(91)90022-T.

Saunders, Phillip M., and David L. VanderZwaag. 2010. "Canada and St. Pierre and Miquelon Transboundary Relations: Battles and Bridges." In *Recasting Transboundary Fisheries Management Arrangements in Light of Sustainability Principles: Canadian and International Perspectives*, edited by Dawn A. Russell and David L. VanderZwaag, 209–38. Leiden: Brill Nijhoff.

Scheufele, Dietram A., and David Tewksbury. 2007. "Framing, Agenda Setting, and Priming: The Evolution of Three Media Effects Models." *Journal of Communication*. https://doi.org/10.1111/j.1460-2466.2006.00326.x.

Schofield, Clive. 2007. "Minding the Gap: The Australia–East Timor Treaty on Certain Maritime Arrangements in the Timor Sea (CMATS)." *International Journal of Marine and Coastal Law* 22 (2): 189–234.

Schofield, Clive. 2008. "Australia's Final Frontiers?: Developments in the Delimitation of Australia's International Maritime Boundaries." *Maritime Studies* 158: 2–21.

Schofield, Clive, and Andreas Østhagen. 2020. "A Divided Arctic: Maritime Boundary Agreements and Disputes in the Arctic Ocean." In *Handbook on Geopolitics and Security in the Arctic*, edited by Joachim Weber, 171–91. Springer.

Schofield, Clive, Rashid Sumaila, and William Cheung. 2016. "Fishing, Not Oil, Is at the Heart of the South China Sea Dispute." *The Conversation*, August 15, 2016.

Schopmans, Hendrik Ruben. 2018. "Explaining (Non-)Cooperation on Disputed Maritime Resources: Joint Development Agreements, Disputed Territory, and Lessons from the Falkland Islands." *Australian Journal of Maritime & Ocean Affairs* 10 (2): 98–117. https://doi.org/10.1080/18366503.2017.1420997.

Scott, James C. 2009. *The Art of Not Being Governed: An Anarchist History of Upland Southeast Asia*. New Haven: Yale University Press.

Shearman, David J. C., and Joseph Wayne Smith. 2007. *The Climate Change Challenge and the Failure of Democracy*. Westport: Praeger.

Shukman, David. 2017. "Renewables' Deep-Sea Mining Conundrum." *BBC News*, April 11, 2017. http://www.bbc.com/news/science-environment-39347620.

Simon, Sheldon W. 2012. "Conflict and Diplomacy in the South China Sea." *Asian Survey* 52 (6): 995–1018.

Siousiouras, P., and G. Chrysochou. 2014. "The Aegean Dispute in the Context of Contemporary Judicial Decisions on Maritime Delimitation." *Laws* 3 (1): 12–49.

Skram, Arild-Inge. 2017. *Alltid til stede: Kystvakten 1997–2017 (Always Present: Coast Guard 1997–2017)*. Bergen: Fagbokforlaget.

Smith, Sheila A. 2012. "Japan and The East China Sea Dispute." *Orbis* 56 (3): 370–90.

Snow, Alpheus Henry. 1913. "International Law and Political Science." *The American Journal of International Law* 7 (2): 315–28.

Solstad, Silje Charlotte. 2012. "Spillet om delelinja: En tonivåanalyse av forhandlingene om grense i Barentshavet og Polhavet (The Game over the Boundary: A Two-Level Analysis of the Negotiations over the Boundary)." University of Tromsø.

Song, Andrew M. 2015. "Pawns, Pirates or Peacemakers: Fishing Boats in the Inter-Korean Maritime Boundary Dispute and Ambivalent Governmentality." *Political Geography* 48: 60–71. https://doi.org/10.1016/j.polgeo.2015.06.002.

Spijkers, Jessica, and Wiebren J. Boonstra. 2017. "Environmental Change and Social Conflict: The Northeast Atlantic Mackerel Dispute." *Regional Environmental Change* 17 (6): 1835–51. https://doi.org/10.1007/s10113-017-1150-4.

Spijkers, Jessica, Tiffany H. Morrison, Robert Blasiak, Graeme S. Cumming, Matthew Osborne, James Watson, and Henrik Österblom. 2018. "Marine Fisheries and Future Ocean Conflict." *Fish and Fisheries* 19 (5): 798–806. https://doi.org/10.1111/faf .12291.

Springer, A. L. 1994. "Do Fences Make Good Neighbours? The Gulf of Maine Revisited." *International Environmental Affairs* 6: 223.

Spruyt, Hendrik. 1994. *The Sovereign State and Its Competitors: An Analysis of Systems Change* Princeton: Princeton University Press.

St-Louis, Carole. 2014. "The Notion of Equity in the Determiniation of Maritime Boundaries and Its Application to the Canada–United States Boundary in the Beaufort Sea." University of Ottawa.

Stabrun, Kristoffer. 2009. "The Grey Zone Agreement of 1978: Fishery Concerns, Security Challerges and Territorial Interests." *FNI Report* 13: 1–43.

Stacey, Natasha. 2007. *Boats To Burn: Bajo Fishing Activity in the Australian Fishing Zone.* Canberra: ANU Press.

Stavridis, James. 2017. *Sea Power: The History and Geopolitics of the World's Oceans.* New York: Penguin Press.

Steinberg, Philip E. 1999. "Navigating to Multiple Horizons: Toward a Geography of Ocean-Space." *Professional Geographer* 51 (3): 366–75.

Steinberg, Philip E. 2001. *The Social Construction of the Ocean.* Cambridge: Cambridge University Press.

Stokke, Olav Schram. 2000. "Managing Straddling Stocks: The Interplay of Global and Regional Regimes." *Ocean and Coastal Management* 43 (2–3): 205–34. https://doi .org/10.1016/S0964-5691(99)00071-X.

Stokke, Olav Schram. 2017. "Geopolitics, Governance, and Arctic Fisheries Politics." In *Global Challenges in the Arctic Region: Sovereignty, Environment and Geopolitical Balance*, edited by E. Conde and S. S. Iglesias, 170–95. London: Routledge.

Stokke, Olav Schram, Lee G. Anderson, and Natalia S. Mirovitskaya. 1999. "The Barents Sea Fisheries." In *The Effectiveness of International Environmental Regimes: Causal Connections and Behavioral Mechanisms*, edited by Oran R. Young, 91–155 Cambridge: MIT Press.

Stokke, Olav Schram, Andreas Østhagen, and Andreas Raspotnik, eds. n.d. *Marine Resources, Climate Change, and International Management Regimes.* London: I. B. Tauris.

Støre, Jonas Gahr 2010. "Norge, Russland Og Delelinjen (Norway, Russia and the Delimitation)." Utenriksdepartementet (Norwegian Ministry of Foreign Affairs). 2010. https://www.regjeringen.no/no/dokumentarkiv/stoltenberg-ii/ud/taler-og-artikler/ 2010/blogg_delelinje/id604680/.

Støre, Jonas Gahr. 2012. "The High North and the Arctic: The Norwegian Perspective." *The Arctic Herald* 2 (June 2012): 8–15. https://www.regjeringen.no/no/aktuelt/nord _arktis/id685072/.

Storey, David. 2012. *Territories: The Claiming of Space.* 2nd ed. Abingdon: Routledge.

Stuart, Jill. 2013. "Regime Theory and the Study of Outer Space Politics." *E-International Relations*, 2013. http://www.e-ir.info/2013/09/10/regime-theory -and-the-study-of-outer-space-politics/.

Supancana, Ida B.R. 2015. "Maritime Boundary Disputes between Indonesia and Malaysia in the Area of Ambalat Block: Some Optional Scenarios for Peaceful Settlement." *Journal of East Asia and International Law* 8 (1): 195–211. https://doi .org/10.14330/jeail.2015.8.1.09.

Swartz, Wilf, Enric Sala, Sean Tracey, Reg Watson, and Daniel Pauly. 2010. "The Spatial Expansion and Ecological Footprint of Fisheries (1950 to Present)." *PLoS ONE* 5 (12). https://doi.org/10.1371/journal.pone.0015143.

Tamnes, Rolf, and Kristine Offerdal. 2014. "Conclusion." In *Geopolitics and Security in the Arctic: Regional Dynamics in a Global World*, edited by Rolf Tamnes and Kristine Offerdal, 166–77. Abingdon: Routledge.

Tan, Andrew T. H. 2007. *The Politics of Maritime Power: A Survey*. London: Routledge.

Tanaka, Yoshifumi. 2013. "Reflections on the Territorial and Maritime Dispute between Nicaragua and Colombia before the International Court of Justice." *Leiden Journal of International Law* 26 (4): 909–31. https://doi.org/10.1017/S0922156513000460.

Tarrow, Sidney. 2001. "Transnational Politics: Contention and Institutions in International Politics." *Annual Review of Political Science* 4 (1): 1–20. https://doi.org/doi:10.1146/annurev.polisci.4.1.1.

Thao, Nguyen Hong, and Ramses Amer. 2007. "Managing Vietnam's Maritime Boundary Disputes." *Ocean Development and International Law* 38 (3): 305–24. https://doi.org/10.1080/00908320701530482.

Theutenberg, Bo Johnson. 1987. "Mare Clausum et Mare Liberum." *Arctic* 37 (4): 481–92.

Thomassen, Ida Cathrine. 2013. "The Continental Shelf of Svalbard: Its Legal Status and the Legal Implications of the Application of the Svalbard Treat Regarding Exploitation of Non-living Resources." University of Tromsø.

Till, Geoffrey. 2004. *Seapower: A Guide for the Twenty-First Century*. London: Frank Cass.

Tiller, Rachel, Elizabeth De Santo, Elizabeth Mendenhall, and Elizabeth Nyman. 2019. "The Once and Future Treaty: Towards a New Regime for Biodiversity in Areas beyond National Jurisdiction." *Marine Policy* 99 (January): 239–42. https://doi.org/10.1016/j.marpol.2018.10.046.

Tilly, Charles. 1990. *Coercion, Capital and European States: AD 990–1990*. 1st ed. Cambridge: Basil Blackwell.

Tovar, Daniel A. 2015. "Colombia and Venezuela: The Border Dispute Over the Gulf." *Council on Hemispheric Affairs*, August 13, 2015.

Treaty of Paris. 1783. "Treaty of Paris 1783." https://www.loc.gov/rr/program/bib/ourdocs/paris.html.

Treglode, Benoit De, and David Buchanan. 2016. "Maritime Boundary Delimitation in the Gulf of Tonkin." *China Perspectives* 3: 33–41.

Tsebelis, George. 1995. "Decision Making in Political Systems: Veto Players in Presidentialism, Parliamentarism, Multicameralism and Multipartyism." *British Journal of Political Science* 25 (3): 289–325.

Tunsjø, Øystein. 2018. *The Return of Bipolarity in World Politics: China, the United States, and Geostructural Realism*. New York: Columbia University Press.

UN General Assembly. 1982. *Convention on the Law of the Sea*. New York: Entry into force: 16 November 1994.

UN Law of the Sea. 2005. *Royal Decree on Amendment of Royal Decree on Delimitation of the Territorial Waters of Greenland, 15 October 2004*. http://www.un.org/Depts/los/doalos_publications/LOSBulletins/bulletinpdf/bulletin56e.pdf.

United Nations. 1946. *Statute of the International Court of Justice*. http://www.icj-cij.org/documents/?p1=4&p2=2.

United Nations. 1958. Convention on the Continental Shelf, Geneva. April 29, 1958.

United Nations. 1982. "United Nations Convention on the Law of the Sea (LOSC)." *1833 U.N.T.S. 397 (in Force 16 November 1994), Publication No. E97.V10.* Montego Bay.

United Nations. 2010. "Treaty between the Kingdom of Norway and the Russian Federation Concerning Maritime Delimitation and Cooperation in the Barents Sea and the Arctic Ocean." http://www.un.org/depts/los/LEGISLATIONAN DTREATIES/PDFFILES/TREATIES/NOR-RUS2010.PDF.

United Nations. 2021. "Status: United Nations Convention on the Law of the Sea." Treaty Collection. New York. 2021. https://treaties.un.org/pages/ViewDetailsIII .aspx?src=TREATY&mtdsg_no=XXI-6&chapter=21&Temp=mtdsg3&clang=_en.

United States. 1945. *Truman Proclamation On The Continental Shelf – Presidential Proclamation No. 2667, 28th September, 1945.* https://iea.uoregon.edu/treaty-text/ 1945-presidentialproclamationnaturalresourcescontinentalshelfentxt.

United States–Panama. 1977. *Panama Canal Treaty of 1977.*

US Department of State. 1995. *Public Notice 2237: Exclusive Economic Zone and Maritime Boundaries.*

US Senate. 1981. "Executive Report No. 5: Maritime Boundary Settlement with Canada." Washington DC.

Valencia-Ospina, E. 1996. "The Use of Chambers of the International Court of Justice." In *Fifty Years of the International Court of Justice: Essays in Honour of Sir Robert Jennings*, edited by V. Lowe and M. Fitzmaurice, 503–27. Cambridge: Cambridge University Press.

VanderZwaag, David L. 2010. "The Gulf of Maine Boundary Dispute and Transboundary Management Challenges: Lessons to be Learned." *Ocean and Coastal Law Journal* 15 (2): 241–60.

Vasquez, John A. 1993. *The War Puzzle.* Cambridge: Cambridge University Press.

Vasquez, John A. 1995. "Why Do Neighbors Fight? Proximity, Interaction, or Territoriality." *Journal of Peace Research* 32 (3): 277–93.

Vasquez, John A., and Brandon Valeriano. 2009. "Territory as a Source of Conflict and a Road to Peace." In *The SAGE Handbook of Conflict Resolution*, edited by Jacob Bercovitch, Victor Kremenyuk, and William Zartman, 193–209. London: Sage.

Vey, Jean-Baptiste. 2019. "France's Macron Sides with Cyprus on Dispute with Turkey." *Reuters*, June 14, 2019. https://www.reuters.com/article/us-eu-france-macron/ frances-macron-sides-with-cyprus-on-dispute-with-turkey-idUSKCN1TF2FO.

Vidas, Davor. 2009. "The UN Convention on the Law of the Sea, the European Union and the Rule of Law: What is Going on in the Adriatic Sea?" *International Journal of Marine and Coastal Law* 24 (1): 1–66. https://doi.org/10.1163/157180808X353902.

Vidas, Davor. 2018. "The Delimitation of the Territorial Sea, the Continental Shelf, and the EEZ: A Comparative Perspective." In *Maritime Boundary Delimitation: The Case Law – Is It Consistent and Predictable?*, edited by Alex G. Oude Elferink, Tore Henriksen, and Signe Veierud Busch, 33–60. Cambridge: Cambridge University Press.

Vogler, John. 2000. *The Global Commons: Environmental and Technological Governance.* 2nd ed. Chichester: John Wiley & Sons.

Walker, Timothy. 2015. "Why Africa Must Resolve its Maritime Boundary Disputes." *Institute for Security Studies* Policy Brief (October): 1–8.

Waltz, Kenneth N. 1959. *Man, the State, and War.* New York: Columbia University Press.

Waltz, Kenneth N. 1979. *Theory of International Politics.* Boston: McGraw-Hill.

Weber, Max. 1946. "Max Weber: Politics as Vocation." In *Max Weber: Essays in Sociology*, 77–128. http://socialpolicy.ucc.ie/Weber_Politics_as_Vocation.htm.

Weil, Prosper. 1989. *The Law of Maritime Delimitation – Reflections*. London: Grotius Publications.

Wendt, Alexander E. 1994. "Collective Identity Formation and the International State." *The American Political Science Review* 88 (2): 384–96.

Wendt, Alexander E. 1999. "Social Theory of International Politics." *American Political Science Review* 94: 429. https://doi.org/10.1017/CBO9780511612183.

Wiegand, Krista E. 2011a. *Enduring Territorial Disputes: Strategies of Bargaining, Coercive Diplomacy, and Settlement. Studies in Security and International Affairs.* Athens: University of Georgia Press.

Wiegand, Krista E. 2011b. "Militarized Territorial Disputes: States' Attempts to Transfer Reputation for Resolve." *Journal of Peace Research* 48 (1): 101–13.

Wiegand, Krista E. 2012. "Bahrain, Qatar, and the Hawar Islands: Resolution of a Gulf Territorial Dispute." *The Middle East Journal* 66 (1): 78–95. https://doi.org/10.3751/66.1.14.

Willheim, Ernst. 1989. "Australia–Indonesia Sea-Bed Boundary Negotiations: Proposals for a Joint Development Zone in the 'Timor Gap.'" *Natural Resources Journal* 29 (Summer): 821.

Wood, B. Dan, and Jeffrey S. Peake. 1998. "The Dynamics of Foreign Policy Agenda Setting." *The American Political Science Review* 92 (1): 173–84.

Wood, Louisa J., Lucy Fish, Josh Laughren, and Daniel Pauly. 2008. "Assessing Progress towards Global Marine Protection Targets: Shortfalls in Information and Action." *Oryx* 42 (3): 340–51. https://doi.org/10.1017/S003060530800046X.

Woody, Todd. 2017. "Seabed Mining: The 30 People Who Could Decide the Fate of the Deep Ocean." *Oceans Deeply*, September 6, 2017. https://www.newsdeeply.com/oceans/articles/2017/09/06/seabed-mining-the-24-people-who-could-decide-the-fate-of-the-deep-ocean.

WTO. 2019. "World Trade Statistical Review 2019." Geneva. https://www.wto.org/english/res_e/statis_e/wts2017_e/wts2017_e.pdf.

Young, Oran R. 1986. "International Regimes: Toward a New Theory of Institutions." *World Politics* 39 (June 2011): 104–22.

Young, Oran R. 1989. *International Cooperation: Building Regimes for Natural Resources and the Environment*. Ithaca: Cornell University Press.

Young, Oran R. 2011. "Effectiveness of International Environmental Regimes: Existing Knowledge, Cutting-Edge Themes, and Research Strategies." *Proceedings of the National Academy of Sciences* 108 (50): 19853–60. http://www.pnas.org/cgi/doi/10.1073/pnas.1111690108.

Young, Oran R. 2012. "Regime Theory Thirty Years On: Taking Stock, Moving Forward." *E-International Relations*, 2012. http://www.e-ir.info/2012/09/18/regime-theory-thirty-years-on-taking-stock-moving-forward/.

Young, Oran R. 2021. *Grand Challenges of Plantary Governance: Global Order in Turbulent Times*. Cheltenham, UK and Northampton, MA, USA: Edward Elgar Publishing.

Zacher, Mark W. 2001. "The Territorial Integrity Norm: International Boundaries and the Use of Force." *International Organization* 55 (2): 215–50.

Zaibang, Wang. 2012. "Approaches to China's Current Maritime Disputes." *Contemporary International Relations* 22 (5): 73–76.

Zyla, Benjamin. 2009. "NATO and Post-Cold War Burden-Sharing: Canada 'the Laggard?'" *International Journal* 64 (2): 337–59.

Index